Radio and Society

Radio and Society:
New Thinking for an Old Medium

Edited by

Matt Mollgaard

Radio and Society: New Thinking for an Old Medium
Edited by Matt Mollgaard

This book first published 2012

Cambridge Scholars Publishing

12 Back Chapman Street, Newcastle upon Tyne, NE6 2XX, UK

British Library Cataloguing in Publication Data
A catalogue record for this book is available from the British Library

ISBN (10): 1-4438-3607-9, ISBN (13): 978-1-4438-3607-4

TABLE OF CONTENTS

ACKNOWLEDGEMENTS

I would like to thank the Faculty of Design and Creative Technologies at Auckland University of Technology for awarding me a research grant to complete this project. I would also like to thank Jennie Watts who did initial editing on most of the chapters in the book. Thank you to J. Mark Percival for good advice on the title. I would like to acknowledge colleagues for their encouragement and support and also all the radio scholars, practitioners and enthusiasts I have got to know over the years; you have always been great company. Finally, thank you to Reina and Hugo for being so patient.

INTRODUCTION

This book features new scholarship for an old medium – radio. For over a century enthusiasts, scholars, practitioners, governments, businesses and listeners have helped to mature the original electronic mass medium – taking it out of the laboratory, into the back shed, out into the world and everywhere people are. There is still no mass medium as ubiquitous as radio, as it permeates our lives in so many ways. Our houses, cars, public spaces and phones all have receivers and we can now hear radio content online too. Count the radio receiving devices around you now; then try to extrapolate from that a rough estimate of all the devices that can receive radio for the entire world. The number is practically infinite anyway, as more receivers, clock radios, cars, mobile phones, computers and other devices that receive radio content are produced across the world every day.

Despite the advent of television, the explosive growth of the Internet, the spread of digital gaming platforms, the revolution in mobile entertainment options and the vast amount of content available in digital databases, radio receivers continue to be built and used. Radio stations are still broadcasting and radio remains a critical part of the media environment. The growth in radio broadcasting in developing nations reminds us that radio is far from being irrelevant and outmoded, as it is still a relatively cheap, easy to make and to access communications technology. In fact, radio has more than survived the critical challenges of the Internet, the computer and digital mobile entertainment; it has co-opted them as new platforms to expand its reach even further. Even the monolithic Apple Inc. has had to include radio receivers in its beautifully designed and cutting edge digital entertainment devices. Notably, the Internet has provided radio operators with a vast new storehouse and transmission platform for their outputs, free of the temporal and geographic constraints of the pre-Internet age.

Radio continues to face critical challenges. The same pressures of commerce and politics, the debates about the utilities of public and commercial broadcasting and ongoing discussions about the power of the media influence radio today - much as they have for the last century.

Radio is still perceived as powerful, influential and capable of creating societal change as well as commercial profits. This tension is being played out in debates about the value of broadcasting to different sections of society against the background of an evolving cornucopia of new media options in the 21st century. The place of radio in this new environment has become contested as never before in that what was understood as radiogenic content—auditory electronic mass media communication—is now able to be recycled, re-imagined and remediated on completely new platforms and devices and without the need or even the desire to broadcast that content.

What "radio" *means* is far more complex than a century ago, and scholarship is reflecting this complexity, exploring the nuances of history, technology, society, commerce, politics and creativity that have been and remain so influential on the medium and its audiences. New ways of talking about radio and what it does and might do are emerging, driven by a desire to understand the reasons why this simple, old and easy media technology is still with us, despite many predictions of its imminent demise during the last 100 years. It can be argued that there is perhaps a sense of common cause, almost a wry sense of satisfaction at times, amongst radio scholars and practitioners at the beginning of the 21st century. This medium that has been regularly challenged by newer technologies and often dismissed as usurped continues to captivate media doers, thinkers and users. It has spread itself into the new digital platforms and has been reinvigorated by them, while adding to them in turn. Radio is now developing synergies with visual aspects of the media, creating hybrid forms of practice and content previously unimaginable. This is another way that radio offers incredibly rich ways to interrogate media forms and how we make and use them, all the while continuing to provide different spheres; large and small, commercial and public, free and fettered, as it always did. Radio is still here, it is still interesting and importantly, it is still being developed.

This book is a collection of contemporary research that explores different aspects of this complex and fascinating media form. The chapters herein cover a broad range of radio topics, from early radio histories to modern developments created by the potentials of the digital age. Several chapters engage with critical debates about the role of government, business and communities in how radio is used in our societies. Other chapters provide important new insights into making radio and radio as a cultural force. Chapters also provide developments in research methodologies that can help us gain further insights into historical and contemporary radio issues. Of course, none of these things exist in isolation from each other and it is hoped that the reader will get a sense of the interwoven influences and potentials of radio; as it has been, is and will be. It is also hoped that this volume provides platforms for more engagement with radio research as a rich, vibrant and fruitful way to further our understandings of the media and ultimately, ourselves.

The book is arranged in a generally chronological and thematic manner, in order to provide different insights into key aspects of the evolution of radio and also to give context to current debates. The book starts with two chapters that explore the earlier years of radio. Peter Hoar provides an account of the challenges that the new technology of "wireless transmission" bought to New Zealand society. This chapter reflects on how the technology of radio was transformative, threatening and ultimately fascinating for those who encountered it in its early stages.

Anne MacLennan further develops these early encounter themes by tracing the links between Canadian newspapers and the growth of radio during the 1930s. The symbiotic relationship that developed between the old and new media forms provides new insights into how they reacted to one another and also how audiences perceived their roles and used them to react to change.

Richard Rudin then takes us to the 1960s and the encroachment of private enterprise on the state radio monopoly in the United Kingdom. This chapter explores how elitist perceptions about commercial radio and its threat to the post-war political, social and cultural consensus in the United Kingdom created conflict and drove detractors of commercial radio to mount a concerted campaign to influence the public and the government away from granting licenses to commercial operators. Here we see the interwoven influences of elites, the press and the government on broadcasting policy, as well as the emergence of another threat to the status quo – "pirate" radio.

Another attempt to break a governmental broadcasting monopoly by "taking to the high seas" is the focus of the next chapter. I argue that pirate radio in New Zealand (in the form of Radio Hauraki), was successful in establishing a new commercial status quo, and that the mystique of pirate radio played a critical part in the lack of debate around the almost total deregulation of broadcasting and the overwhelming commercial nature of radio in New Zealand. This chapter interrogates some of the myths and effects of the pirate broadcasting era and shows the very different outcomes of the pirate radio challenge for New Zealand, despite similarities with the experiences of the United Kingdom that Rudin explores.

Tom Morton takes us to Australia and the development of an Australian music scene, which is fostered by Australian radio from the 1970s on. This is confronted and explored in a series of radio documentaries, structured as a cultural history of the "Australian sound". Here we see the intersections between music and radio, but also the contemporary potential of sound broadcasting as storytelling and remembering. This chapter also interrogates narratives and ways of mapping the past as well as how radio can help reinvigorate that history for audiences now.

Sam Coley discusses music and radio documentary too, while also exploring ideas of fandom and how fans use the Internet to repurpose collectable material and to display their devotion. By taking us from the height of David Bowie's musical career, to discovering a previously unheard Bowie song and then to documentaries made 25 years later and remixed by Bowie fans, Coley interrogates notions of fandom and also radio content on the Internet, providing us with insights into how the Internet can expand radio's potential, but also how audiences can re-imagine radio content, given the right tools and motivations.

The next three chapters take us into the contemporary radio station to discuss what influences are acting on what we hear. Harry Criticos questions what is lost when a combination of government policies, ownership changes, commercial pressures and technology allows radio companies to network their stations together, losing local content and input. This chapter reveals some of the disadvantages of losing local radio influences, programmes and workers, as well as discussing the commercial pressures that privilege regional and metropolitan content over localness in the contemporary radio environment.

J. Mark Percival also takes us behind the scenes by exploring what makes "good" radio music. Radio music programmers have a profound influence on popular culture and their decisions can make or break entire careers or even genres. Percival talks to the programmers to find out how they imagine they make these critical decisions and critiques their approaches to choosing the songs they allow on their radio stations. These practices are virtually invisible to listeners and musicians; this chapter is a unique insight into how music gets selected for airplay on the radio.

Helen Wolfenden grapples with one of the great clichés of radio performance, the constant advice to "just be yourself". This chapter explores one of the great dilemmas that face radio broadcasters, especially new ones: how to be one's self, while performing. Wolfenden discusses critical elements of finding and selecting selves to be, being "authentic" and what performance for the audience means to radio broadcasters. This chapter reveals the complex nature of what is actually happening for the broadcaster as a person and a performer when the microphone is turned on.

Tony Stoller takes us into the halls of government. Stoller is an insider, having been intricately involved with the development and deployment of digital radio in the United Kingdom. Here he reveals how broadcasting policy can be made with almost no deep critical oversight and also how technological fervour can create unexpected outcomes for governments, broadcasters and listeners. This chapter can be read as a somewhat cautionary tale, complete with insider insights and nuanced observations of how policy makers can be enraptured, and even captured, by new technologies and the desire to be seen to be embracing them.

The next chapter is also from an insider. Brent Simpson has spent many years working in the Low Power FM scene in New Zealand and is a strong advocate for community broadcasting. Simpson explores the open-access Low Power FM system in New Zealand, which is unique in the world. Using an "Open Commons" approach, Simpson asks if the utility of Low Power FM is being fully realized in New Zealand and critiques successive government's reluctance to reinforce the sector, arguing that it has demonstrated potential to do more good for New Zealand citizens. This chapter challenges policy makers to utilise broadcasting legislation to engage with the LPFM sector more fully.

Matt Grimes and Siobhan Stevenson give us insights into how community-driven radio can help to rehabilitate and involve individuals and groups in society. Their research reveals the power of radio to help those on the margins empower themselves and others through their stories and through working together to produce them, as well as helping them pick up transferrable skills. Working with prisoners and the Travelling community, Grimes and Stevenson help us to understand the power of radio to make those that are ignored, forgotten or despised feel empowered, listened to and part of a wider society. Their research methods provide new ways of designing projects to work with those who would greatly benefit from being acknowledged and supported in telling their stories.

Janey Gordon also explores research methods in the next chapter. Gordon discusses approaches to researching the audiences of community broadcasters in an inexpensive but in-depth and valid way. This chapter discusses qualitative and quantitative methods for gaining good information about radio audiences that can be used to fine tune programming and services. The chapter includes the results of a recent audience research project at a community radio station in the United Kingdom and how these methods could be used by others seeking reliable insights into their audiences at minimal cost.

The final chapter looks at how a large radio company is adapting to the realities of the Internet age. Pierre C. Bélanger explores how a major Canadian radio broadcaster is re-orientating itself towards the web and how that affects what people in the organisation do in order to participate in a "dynamic web strategy", which goes beyond the broadcast and into the personal spaces of the web and increasingly, Internet-connected mobile phones and computing. What is revealed is that the business model for traditional broadcasters is rapidly changing, and that work practices, content creation and also delivery should reflect this. Bélanger puts forward a strong case for developing the human and technological resources to reach increasingly distracted audiences in the spaces and on the platforms they are using now and will be using in the future.

These chapters represent a small but significant collection of the increasing amount of new work being generated by scholars interested in the longevity, ubiquity, utility and flexibility of radio. There is much more to be done for sure, but it is hoped that this volume gives the reader a sense of the richness and potential of radio studies, as well as the passion for extending understandings of the medium held by many scholars of the media today.

CHAPTER ONE

MORSE, MAGIC AND MODERNITY: RECEIVING RADIO IN NEW ZEALAND 1900–1914

PETER HOAR

Radio is a medium that is not so much forgotten as taken for granted. It is always there, always on, in our cars, in our houses and on our iPods. The medium's mixture of voices, music and often jarring advertisements is part of the background noise of contemporary life. It may seem obvious what "radio" is, what it does, and what it means.

But our modern understanding of the term can be problematic when we try and understand the history of the medium, let alone any other audio technology such as phonographs, telephones or film. We tend to hear these audio technologies through the noise and interference of our modern meanings of the terms used to name them. As Marvin pointed out, "Media are not fixed objects: they have no natural edges" (1988, p. 8).

Our modern understandings of words such as "radio", "recording" or "film" are not those of the past. In New Zealand, as elsewhere, early "radio" was understood as wireless telegraphy i.e. an extension of the sort of single point to single point communication used in telegraphic communications, or letter writing. These days, radio is usually assumed to be one-to-many transmission, or broadcasting. Radio transmissions before the 1920s were in Morse code and heard through earphones. The earphones may remain today in some cases, although most modern radio listening actually happens in cars, but Morse code is not what we hear today. Modern radio broadcasting involves an unbroken wall of sound (speech, music, sound effects) designed to keep listeners ready for the commercial messages that fund most radio broadcasting in New Zealand. But the sounds of radio in New Zealand between 1900 and 1914 were the sparse and sporadic "dits" and "dahs" of tapped out Morse messages set against, and often lost in, the whooshing, buzzing, clicking, howling, humming, sometimes screaming, and always present sounds of electromagnetic waves interacting in the Earth's atmosphere. The heavenly harmonies of the music of the spheres turned out to be static.

Calling this audio experience "listening to radio" seems to muffle the historical experiences involved and deafens us to the cultural meanings and significances that these media once had. This highlights what Wurtzler pointed out: "The distinctions between media forms were not as clear as media–specific historical scholarship might suggest" (2007, p. 11). This historical enquiry in turn destabilizes our use of terms for modern media such as the "Internet" or "mobile phone" and problematises current debates about media and technologies (Hendy, 2008; Lacey, 2008).

Debates about media and technology are lacking in New Zealand historiography (Smithies, 2005-2006). General histories make token references to technologies such as refrigerated shipping when these are involved with larger stories about agriculture and economic change. Media such as radio, television and cinema are dutifully noted as important parts of social and cultural life but the significant roles technologies have played in New Zealand's past tend to be downplayed (Belich, 2001; King, 2003; Smith, 2005). Specialised histories about New Zealand's media have concentrated on institutional developments. These have often tended to have a nationalist focus built on teleological foundations. Their narratives describe a progress from isolation, obscurity and simple origins through to a triumphant "now" where New Zealand film, music, radio, and other cultural products are as good as any in the world and proudly reflect specifically local conditions and circumstances (Day, 1994, 2000; Staff & Ashley, 2002; Pivac, Stark, McDonald, 2011).

Thinking about radio in New Zealand before 1914 undermines such certainties about progress and isolation and also blurs the edges of the word "radio" itself. Radio places New Zealand squarely in the middle of modernity. The ways in which people learned about radio and understood it also decentre New Zealand as a subject. National history, conceived and inscribed as national identity, as an organic and normative formation, is in itself a practice of colonization. Teleological and nationalist histories tend to reinforce this cultural colonization. By thinking about the world in New Zealand, rather than New Zealand in the world, it may be possible to undermine these normative historiographical practices. Cultural histories about media and technology are particularly suited to such a decentring process and this chapter is an attempt to do this through some of the meanings that radio had in New Zealand before 1914 (Gibbons, 2002, 2003; Byrnes, 2010). One important idea about radio in those days was that it was a threat to New Zealand's security as Eric Battershill discovered in 1913 when he found himself in court facing serious charges.

Battershill was a schoolboy living in Hastings, a small town in New Zealand's North Island, and he had a great interest in electronics and a great skill for tinkering with gadgets. By combining these talents and with some technical information from books and magazines, he built from scratch a radio set that could receive signals from thirty kilometres away but could not transmit. He was a typical amateur radio user in the years before the Great War. Typical in that he was a boy. The only woman we know of being directly involved with New Zealand radio in this period was Margaret Bell. She, with her brother Francis, operated a radio set at the Shag Valley sheep station deep in the South Island province of Otago (Dougherty, 2007). But the majority of radio amateurs were male and they were often schoolboys, such as Eric, who were following their interests in science. Tinkering with the audio technologies of modernity, such as phonographs and radios was a male dominated activity, both in New Zealand and in other countries where experiments were being made with radio transmission (Douglas, 1987, pp.190-92). Ingenious a tinkerer as he was, Eric found himself in the Hastings court as his radio activities were illegal. He had no licence to operate his set. Radio in 1913 was very much a matter of state control.

New Zealand had been the first country in the world to enshrine legislation controlling the radio medium. The New Zealand Wireless Telegraphy Act, passed on 26 Spetember1903, predated the British Wireless Telegraph Act of 1904. The New Zealand Act was designed to establish a government monopoly over the nascent medium and it imposed draconian penalties on those who breached it. Unlicensed operators faced their confiscation of their laboriously constructed equipment as well as a possible £500 fine (Wilson, 1994, p. 92). New Zealand depended on shipping for its survival and the government of the day understood the role that wireless telegraphy could play in safeguarding marine trade. There was a precedent for government ownership of communication technologies in the form of the telegraph system and the expanding telephone network that were both managed by the New Zealand Post Office. The Act addressed concerns about privacy in the new realm of wireless communication. The Postmaster-General, Sir Joseph Ward, spoke of the insecurity of "Marconigrams" that circled in the ether until some unscrupulous or perhaps curious private person intercepted them instead of the intended recipient (Day, 1994, p.14). Not only was personal privacy at stake, but so also were important matters of state and Empire. Thus the New Zealand government began the slow business of establishing a chain of radio stations with the first being established in Wellington in 1911. By 1914, there was a systematically designed chain of these stations that provided for wireless telegraphs to be sent overseas as well as maintaining communications with the ships that were the basis of the country's economy (Wilson, pp. 95-96). Wireless telegraphy was a communication medium such as the telephone and telegraph system that was maintained by New Zealand government for the overall good of the country.

Amateur and non-official radio operators were shut off from the airwaves, not only by the 1903 Act but also by the Post and Telegraph Amendment Act of 1913. This latter Act was even stricter than the first in that it made it an offence to construct a device even capable of sending and receiving radio signals. The 1903 Act had criminalised act of unauthorised radio transmission or reception but the 1913 Act penalised even the potential for such activity. Harry Bell, Minister of Internal Affairs, made the point of this legislation loud and clear when he told the Legislative Council that "Private wireless stations are a nuisance wherever they are; and frankly, it is not intended to permit them" (Dougherty, 1997, p.21). Up against this rigid system of carefully controlled and disciplined wireless telegraphy, it might seem that young Battershill was in deep trouble.

The prosecution in Battershill's trial raised a point that had been an important part of the drive to place wireless telegraphy under governmental supervision. It was a kind of slippery slope argument but it certainly seemed to highlight what was at stake in the trial. It was the security of the realm; what Battershill was "doing for his own instruction might be done by others for reasons serious to the country and Empire such as the interception of wireless messages from overseas. Such plots might produce results of the utmost gravity" ("Boy and His Wireless Apparatus", 1913, p. 10). This theme of national and Imperial defence had been taken up by a New Zealand newspaper article in 1912 which had pointed out the vulnerability of the telegraph system to attack and called for all British possessions to be in wireless contact and so form "an enduring line of defence between the Empire and the outer world" ("A Girdle of Wireless", 1912, p. 6). According to the official view, Battershill seemed to be threat to national security rather than a backyard hobbyist, tinkering with the latest communication technology.

The defence scoffed at this idea and maintained that the wireless regulation legislation served only to stultify the talent of scientifically minded youths such as Battershill. No harm had been done and he had been encouraged in his technical endeavours by his teachers at Napier High School. In the event, he was discharged on condition of not using the apparatus again.   Eric Battershill seems to fade out of the radiophonic history of New Zealand after his trial. Whether he carried on experimenting with radio is unknown. His trial is of note in that he was charged in the first place and that the threat posed by wireless telegraphy was being taken seriously at a time when international tensions were heightening due to the naval arms race between Great Britain and Germany. Previously, the wireless regulations were hardly enforced at all.

Sometimes, unofficial and unlicensed wireless activities received official recognition and commendation despite the flouting of the draconian 1903 Act. This was the case of the so-called "three clever boys" in 1908 who were operating illegal wireless equipment at Dunedin, in the South Island ("Three Clever Boys". 1908). Far from being taken to court and being threatened with large fines, these young wireless operators were rewarded with a notice of congratulations from the Premier Joseph Ward. The story behind their achievement illustrates how many early wireless experimenters began working with the new technology.

Rawson Stark, Stanton Hicks and Cyril Brandon were in their late teens when they publically demonstrated wireless transmission on 10 September 1908. They sent messages back and forth across the Otago Harbour including greetings between the mayors of West Harbour and Dunedin along with a message to the Premier which was forwarded to Wellington by telegraph. A number of public figures and officials were present at this demonstration and it was reported at length in local and national newspapers (Day, p.19; Dougherty, pp.15-16).

This demonstration came after two years of study and hard work by the boys. Stark's father was employed as one of the city's electrical engineers while Brandon worked for a private electrical engineering company. They spent all their pocket money on the parts needed to build their equipment from scratch. What they could not buy, they salvaged and scrounged from the city's electrical workshops. Their family connections may have been useful here. Getting hold of the materials for their experiments was one thing but finding out how to put it all together and make it work was another. They spent many hours in the Dunedin library "in search of the latest literature relating to wireless phenomena" and talked to local Post Office telegraph operators and engineers ("Three Clever Boys", 1908). It took a lot of experimentation and effort in building and rebuilding their apparatus before they began to achieve results. After their spectacular demonstration, the mayor of Dunedin expressed his pride that the first land based wireless telegraphy system in New Zealand had been established in the city. Other officials offered similar encomiums to the boys. No mention was made of the illegality of the experiments let alone the threat of prosecution.

The "three clever boys" of Dunedin were rewarded by officials in a spirit of avuncular benevolence that was notably lacking just five years later when Eric Battershill was taken to court. His case showed some of the anxieties and fears associated with the thoroughly modern technology of radio while the Dunedin demonstrations were illuminated with the glamour and fascination that this seemingly miraculous scientific development radiated along with its invisible electromagnetic waves. It was not just technically inclined schoolboys, Post Office technicians and engineers who were interested in radio and informed about it. The wider public were kept up to date about this latest technology along with other global developments. The ways in which knowledge about radio was spread throughout New Zealand illustrates how modernity was experienced at what seemed to be one of the remotest parts of the globe.

The 2 January 1907 issue of the New Zealand popular scientific magazine *Progress* carried a long article by one Captain Louis E. Walker. Walker was an agent of the Marconi Company and he was trying to interest the New Zealand government in purchasing equipment for the proposed network of wireless stations. Walker's article gave a history of wireless development to date or at least the triumphs of the Marconi Company in this field. Along with detailed descriptions of the technology and theories behind radio, the article also included various illustrations. One of these was of the Marconi wireless apparatus on display at the New Zealand International Exhibition that was held in Christchurch 1906-7. Another one was of a circuit diagram of the "Marconi transmitting and receiving apparatus" (Walker, 1907, pp. 92-94). These illustrations can illuminate some aspects of the reception of modernity in New Zealand.

The Marconi Company set up a radio station at the Exhibition as part of its strategy to corner the wireless telegraphy market in Australasia. This was the first public display of wireless in New Zealand and the regular demonstrations sparked off a great interest in the new technology (Day, 1994, pp.16-17). The Exhibition itself was an important moment in New Zealand history in that it marked the country's transition from Colony to Dominion status. Through the Exhibition, New Zealand was displayed to the world, and itself, as a prosperous, advanced and modern country (Phillips, 1998). But the world was also being brought to New Zealand as many countries had displays as part of the Exhibition. The international nature of the Exhibition made it more than a nationalistic statement about New Zealand's role in the world. It was also connecting the rest of the world to New Zealand and so undermining the notion of the new Dominion as an isolated and peripheral island fastness, remote from the distant world centres of North America and Europe. It was showing that New Zealand was part of the modern world and that world included technologies such as wireless telegraphy.

Walker's article also featured a clear and detailed circuit diagram (Walker, 1907, p. 94). It would have been a simple matter to build a duplicate of Marconi's machine from this diagram and the descriptions in the article. For wireless devotees such as Battershill and the "three clever boys", there was a wealth of information available from overseas sources. Similar articles and diagrams were found in magazines such as *Popular Mechanics* and newspapers also published articles that explained the technical aspects of wireless in some detail. Manuals, handbooks and the equivalents of modern "How to" guides were readily available from local bookshops or could easily be bought from overseas. Oliver Lodge's *Signalling Through Space Without Wires* (London, 1900), George Pierce's *Principles of Wireless Telegraphy* (New York, 1910) and C.C.F. Monckton's *Radio–Telegraphy* (London, 1908) were the sorts of works consulted by New Zealand radio enthusiasts for circuit diagrams, equations and general technical advice.

However, most New Zealanders would have received radio through the pages of newspapers and popular magazines rather than technical manuals and specialised journals. Editorials, articles, pictures and cartoons described radio and made it familiar to New Zealanders long before they had a chance to hear it. Most people would have got radio long before they received it. And to get radio, as in to understand or get other technologies such as phonographs, films and telephones, was to be modern. It was to be familiar with the progress of science, the annihilation of time and space, the shrinking of the world, the increasing, exciting and nerve tingling acceleration of the speed of modern life: to know about and maybe use these things was to be part of modernity and New Zealand was a country that was "born modern" in that it entered and participated in the modern world of technology, mass media, and consumption of industrially produced global culture from the genesis of its awareness of itself as a country called "New Zealand". This could, be dated to the 1850s with the establishment of the country as a nation-state (Daley, 2010).

While there may have been few radio waves in the air in pre-1914 New Zealand, the idea of radio was certainly in the air for most people. New Zealand's media had been connected to the global network of telegraph lines since 1876 and the world's news, fashions, fads and technological developments were all part of the daily life of the country's people. Accounts of the experiments and inventions by people such as Thomas Edison and Marconi had appeared in New Zealand newspapers fresh from the European and American presses and telegraph wires. The idea of wireless transmission was familiar to many New Zealanders well before the technology itself. This may have given rise to tensions and anxieties about national security but other meanings were more benevolent. Part of being modern was a form of nonchalance in the face of seemingly miraculous technologies such as wireless communication. Telegraphs, telephones, recorded sound and films were all accepted parts of New Zealand's daily life by the time wireless was demonstrated at the 1906-7 Exhibition. Cartoons and jokes, along with sober technical accounts and demonstrations, played important roles in the normalisation of technology. To get the joke meant that you got the technology even if you did not actually have it as such.

A cartoon from the *New Zealand Free Lance* in 1907 showed how knowledge of wireless communication could be used. The cartoon showed a drunken man woozily clinging for support to a telegraph pole on Lambton Quay, a main street in New Zealand's capital city, Wellington. The caption read: "Don't talk – hic – to me – of new inventions. What'll I do – hic - when they go in - hic - for this new-fangled - hic - wireless telegraphy?" ("Too Much Easter", 1907, p. 15). What this cartoon played on was its audience's knowledge that wireless telegraphy was ... wireless! Unlike the old system of telegraphy, wireless telegraphy used no wires so it needed no poles for these non-existent wires which further meant that the drunks of Lambton Quay would have nothing to cling to when this new technology was fully deployed. The knowing readers of this cartoon, all citizens of a modern country that was deeply embedded in the world of modernity, knew that wireless did indeed mean no wires. This was one of the important things about this new technology that people knew even before they saw it. This familiarity helped demystify the technology itself.

The speed of wireless telegraphy was also made fun of. Another cartoon featured in the *New Zealand Free Lance* in 1908 showed two tramps staring in amazement after a blurred shape that had just zoomed past them at immense speed, causing one to stagger and lose his hat. One tramp wondered if it was a streak of lightening or a "wireless telegram". The other put him right by saying that it was Joseph Ward on the election trail ("Sir Joe in a Hurry", 1908, p. 12). Ward's 1908 campaign saw him tour a lot of the country at high speed. This was not the funniest political cartoon of modern times but it said something about how radio was understood. Radio was fast, like lightening, or a campaigning politician. People knew that radio waves were invisible so part of the knowing humour lay in the absurd idea that one might see a wireless telegraph message. The clued up second tramp got this. So did the readers. But the speed of radio was the idea being played with in this cartoon. Everyone understood that radio was really, really fast.

There was a further political layer to this cartoon in that Ward was very interested in the development of radio and strove for New Zealand to be part of the invisible empire of wireless networks that spread around the world before 1914. Ward was Postmaster-General in 1905-6 when he toured Europe and the USA and attended many demonstrations by various wireless companies. He became familiar with the technology and understood its strategic and commercial possibilities. After becoming Premier in the 1908 election, Ward pushed along the development of official wireless communications in New Zealand (Day, 1994, pp.15-19; Wilson, 1994, pp.91-96). The message of congratulations he sent to the Dunedin boys after their 1912 demonstration was more than a mere token. Ward had a real interest in developing wireless communications in New Zealand. Hence the comparison of him to a wireless telegraph message had multiple and interrelated meanings.

But these cartoons, and many others, along with articles and features in the general media, were ways in which those who were not radio "geeks", such as Eric Battershill or the three clever boys, came to get radio before they actually received it. They were informed citizens of modernity who enjoyed the thrills of new technologies along with the senses of progress and improvement these new gadgets brought in their wake.

The few citizens who got radio in the sense of actually hearing it were engaging in a modern form of listening. The accounts left by New Zealand's wireless pioneers are sparse and tend to concentrate on technical issues and evading detection. These records tend to leave out what was actually heard.

One account from 1922 gives some idea of what radio listening may have been like in the pre-1914 era. When Ken Collins, Chief Technician at the government owned radio station *2YA* until 1949, recalled his first experience of radio, he described hearing through earphones a "faint high-pitched musical note in sharp staccato Morse code – "dit dit dit *dah*, dit dit dit dit". Suddenly the singing note died and immediately, much louder, a dry tearing sound took up the Morse code" (Collins, 1967, p. 13). These sounds of a Sydney radio station communicating with a ship in the Tasman Sea were heard through the inner sonic space generated by earphones.

This personalized, interior sonic space was typical of listening to early acoustic technologies in modernity. Phonographs were often heard through earphones and the telephone had accustomed people to the idea of "sounds in their ears" in new and intimate ways. In fact, many of the early radio listeners had to make their own earphones (along with all other parts of the apparatus) before the mid 1920s, when loudspeakers became more widely available. Telephones could be cannibalized to make earphones. One early radio user recalled that "The phones were of P. and T. [Postal and Telegraph] pattern, obtained from never mind where, suffice to say that they were obtained" (Spackman, 1932, p. 2). Far from being a communal or family experience, radio before the mid-1920s was a fundamentally individual one. The earphones would be passed around if several people were present or multiple sets could be used but each listener heard the sounds in a more intimate manner than if they issued from speakers. The sounds were "in their heads" in a way other sounds were not.

Modern discussions about the use of earphones as listening devices have tended to concentrate on the ways in which they create private aesthetic spaces literally within the listener's head (Bull, 2000, pp. 31-42, 156-161; Weber, 2010). Such accounts stress the novelty of the sort of interior spaces created by earphones but their modern focus overlooks the fact that early audio technologies also frequently involved the use of earphones. Early phonographs, radios and films were often heard through earphones. Their listeners might well have experienced the same kind of inner musical spaces that scholars now associate with the iPod. The inner acoustic space created by later devices such as the Walkman or iPod is less of an innovation than we might think. Modern listening experiences have surprisingly long, complex and varied genealogies that are not purely dependent on reaching some absolute level of technological progress.

Collins described his sense of wonder and the thrill of even hearing these sounds as he could not understand Morse code itself. That point for many of the early radio operators was simply building the equipment, making it work, and hearing something, anything, that could be a radio signal. The goal then became to build more effective equipment and transmit and receive signals over greater distances through this seemingly magical medium that could transcend space and send messages instantaneously.

Susan Douglas has called attention to the way radio appeared as magical and unearthly to its early listeners. In her account, the communication of live people through early radio "bridged the widening gap between machines and spirituality. Radio burrowed into this unspoken longing for contact with the heavens, for a more perfect community, for a spiritual transcendence not at odds with, but made possible by, machines" (2004, p. 41).

New Zealand's early radio operators seem to have been more concerned with technical problems and static rather than bridging the alleged gap between technology and the spiritual. However, the response of listeners such as Collins shows how they were moved and excited by the re-embodiment of human presence at a distance through this new and exciting machinery of modernity. Others did experience wireless as a form of spiritual communication.

The links between wireless communication and discourses around the paranormal have been noted by many scholars (Peters, 1999: Sconce, 2000; Enns, 2006). The international interest in spiritualism was felt in New Zealand with many famous mediums and speakers touring the country as well home grown attempts to communicate with the dead (Ellwood, 1993). Audio technologies such as the telephone and the phonograph had detached the voice from the body and conquered space and time. The ability of the phonograph to capture, preserve and replay voices seemed to even conquer death. Wireless apparatus, radio waves and the ether provided metaphors, and at times mechanisms, for communication with the dead.

New Zealand's spiritualists, devotees of telepathy, ghost hunters and even astrologers were quick to seize on wireless telegraphy as a model for their modes of other worldly communication. Hence a 1903 *Wanganui Herald* article described telepathy as "the wireless telegraphy of the mind" ("Converts to Telepathy", p. 6). In 1906, Nelson astrologer Joseph Taylor, compared finding an individual's "cosmic key" to radio tuning (Taylor, p. 1). The Theosophist Annie Besant, when interviewed during a 1908 lecture tour of New Zealand, spoke of the power of thought as a natural force like wireless telegraphy and "most potent in its effects when rightly understood and directed" ("Mrs Besant's Visit", p. 14). These ways of using the idea of radio might seem like attempts at bridging Susan Douglas's gap between the spiritual and the mechanical mentioned above. If spiritualism is like radio and yet it lets us communicate with the dead then surely some of the otherworldly aspects of spiritualism rub off on radio? The technology itself seems like magic in just the same way talking with the dead is magical and spiritually transcendent. But the transcendence is not so much made possible by the machines as that the technology becomes a quasi-scientific explanatory factor for transcendence itself. The new acoustic technologies were heard by some as ways of explaining beliefs and practices that did not have too much purchase in terms of post-Enlightenment scientific rationality. The howling, buzzing, screaming and sputtering sounds that often drowned out the Morse code messages of wireless telegraphy were heard by some as the real messages being transmitted via the new and non-human medium.

The outbreak of war in 1914 put paid to any further public amateur radio experimentation in New Zealand. Any attempts at unauthorised wireless transmission were treated very seriously. One man was imprisoned early in the war and several others fined heavily for illegal radio ownership and operation. But even so, some amateurs still operated clandestinely during the war years. A group in Gisborne managed to pick up signals from Apia in Western Samoa with a crystal set and a 90 metre aerial that was concealed in a pine tree during the day and unrolled at night (Day, 1994, pp. 30-32; Dougherty, pp. 21-22). There was also a great demand by the government for radio operators in both the navy and the army and as trainers for other operators. The first military action by New

Zealand in the Great War came when a force was sent to Apia to take control of a German radio station established there. The signals received by the Gisborne operators at night were those of the New Zealand garrison established on Samoa to protect this important strategic asset. After the war, a seasoned cadre of radio operators returned to New Zealand and most began experimenting with broadcasting voices and music ("wireless telephony") rather than the one-to-one Morse code transmissions of the pre-war period. Some amateurs still kept to the idea of point to point transmission and developed the DX or ham radio culture that still runs alongside New Zealand's modern broadcast radio market.

Much of the history of New Zealand's media and media technologies remains obscure and many accounts take nationalist and teleological perspectives that can obscure the ways in which the wider world was experienced. A tendency to see New Zealanders as agricultural producers rather than metropolitan consumers of global culture has often led to stories of development that end in a triumphant "now" with plucky New Zealand taking its turn on the world stage with its unique home brewed cultural products that reflect a distinctive sense of national identity.

Reflecting on the ways in which the global networks of technological and cultural modernity were experienced is a way of capturing a snapshot of cultural colonization in action. This process is ongoing and constantly re-inscribed by the very nationalist based histories, films, music and other cultural products that ignore or downplay the role of global influences and corporate entities that shaped the culture of modernity in New Zealand as much as they made cultures in the Old and New worlds. Stories about early radio in New Zealand decentre the nation as the subject and complicate ideas about cultural colonization.

They also destabilise our notions about media and blur the edges of our idea of radio as heard through its history. It may also be that by listening to the static and interference ridden past of radio, and its reception as a new medium then, we might in turn open up new modes of thinking about the reception of new media now.

References

Belich, J. (2001), *Paradise reforged: A history of the New Zealanders.* Auckland: Allen Lane.

Boy and His Wireless Apparatus. (August 14 1913), *Evening Post*, p. 10.

Bull, M. (2000). *Sounding out the city: Personal stereos and the management of everyday life.* Oxford: Berg.

Byrnes, G. (2009). Introduction: reframing New Zealand history. In G. Byrnes (ed.), *The new Oxford history of New Zealand* (pp.1-18). Melbourne, Australia: Oxford University Press.

Collins, G.C. (1967). *Broadcasting grave and gay.* Christchurch: Caxton Press.

Converts to Telepathy. (2 June 1903), *Wanganui Herald*, p. 6.

Daley, C. (2009). Modernity, consumption and leisure. In G. Byrnes (ed.), *The new Oxford history of New Zealand* (pp.423-446). Melbourne, Australia: Oxford University Press.

Day, P. (1994; 2000). A history of broadcasting in New Zealand. Vol. 2., Auckland: Auckland University Press.

Dougherty, I. (1997). *Ham shacks & rag chewers: A history of amateur radio in New Zealand.* Wellington: Historical Branch, Department of Internal Affairs.

—. (2007). Bell, Francis Wirgman Dillon 1896-1987; Bell, Margaret Brenda 1891-1979. *Dictionary of New Zealand Biography.* Retrieved September 1, 2010, from http://www.dnzb.govt.nz/

Douglas, S.J. (1987). *Inventing American broadcasting 1899-1922.* Baltimore: John Hopkins University Press.

—. (2004). *Listening in: Radio and the American imagination.* Minneapolis: University of Minneapolis Press.

Ellwood, R.S. (1993) *Islands of the dawn: The story of alternative spirituality in New Zealand.* Honolulu: University of Hawaii Press.

Enns, A. (2006). Psychic radio: sound technologies, ether bodies and spiritual vibrations. *Senses & Society*, 3, 2, pp.137-152.

Gibbons, P. (2002). Cultural colonization and national identity. *The New Zealand Journal of History*, 36, 1, pp. 5-17.

—. (2003). The far side of the search for identity: reconsidering New Zealand history. *The New Zealand Journal of History*, 37, 1, pp. 38-49.

A Girdle of Wireless. (15 March 1912), *Evening Post*, p. 6.

Hendy, D. (2008). Radio's cultural turns. *Cinema Journal*, 48, 1, pp. 130-138.

King, M. (2003), *The Penguin history of New Zealand.* Auckland: Penguin.

Lacey, K. (2008). Ten years of radio studies: The very idea. *The Radio Journal – International Studies in Broadcast and Audio Media*, 6, 1, pp. 21-31.

Marvin, C. (1988). *When old technologies were new*. New York: Oxford University Press.

Mrs Besant's Visit (19 August 1908), *Otago Witness*, p. 14.

Peters, J.D. (1999). *Speaking into the air*. Chicago: University of Chicago Press.

Phillips, J. (1998). Exhibiting ourselves: The Exhibition and national identity.   In J.M. Thomson (ed.), *Farewell colonialism: The New Zealand Exhibition Christchurch 1906-07* (pp. 17-26). Palmerston North: Dunmore Press.

Pivac, D., Stark, F., McDonald, L. D. (Eds). (2011). *New Zealand film: An illustrated history*. Wellington: Te Papa Press.

Sconce, J. (2000). *Haunted media: Electronic presence from telegraphy to television*. Durham: Duke University Press.

Sir Joe in a Hurry. (7 November 1908), *New Zealand Free Lance*. p. 12.

Smith, P.M. (2005). *A Concise history of New Zealand*. Cambridge: Cambridge University Press.

Smithies, J. (2005-2006). The history of technology and the history of New Zealand. *Journal of New Zealand Studies*, 4-5, pp. 111-128.

Spackman, L.S.S. (1932). When radio was very young: The story of New Zealand radio in the pre-Company Days. *New Zealand Radio Times*, 1 April. pp. 1-2.

Staff, B., Ashley, S. (2002). *For the record: A history of the recording industry in New Zealand*. Auckland: David Bateman.

Taylor, J. (30 October 1906). Principles of astrology. *Nelson Evening Mail*, p. 1.

Three Clever Boys. (16 September 1908), *Otago Witness*, p. 66.

Too Much Easter. (6 April 1907), *New Zealand Free Lance*, p.15.

Walker, L.E. (2 January 1907). Wireless telegraphy, *Progress*, pp. 92-94.

Weber, H. (2010). Head cocoons: A sensori-social history of earphone use in West Germany 1950-2010, *Senses & Society*, 5, 3, pp. 339-363.

Wilson, A.C. (1994). *Wire & wireless: A history of telecommunications in New Zealand 1890-1987*. Palmerston North: Dunmore Press.

Wurtzler, S. (2007). *Electric sounds: Technological change and the rise of corporate mass media*. New York: Columbia University Press.

Chapter Two

Reading Radio: The Intersection Between Radio and Newspapers for the Canadian Radio Listener in the 1930s

Anne F. MacLennan

Introduction

Just as radio was a new medium in the 1930s, so too was the audience. The Canadian radio listeners of the thirties were enthusiastic consumers of the technology who learned about radio, primarily, from its broadcasts. But a second site for radio listeners to learn about this new medium was the local newspaper. Listening norms, national radio strategies, and programming likes and dislikes were all a part of standard radio columns. Whether or not the columns voiced or influenced opinion about radio, they provided a constant stream of information about the technology, programme content, and practice of listening to the radio. So the experience of listening to radio created an audience in the first instance, but the newspaper played a significant role in reinforcing that listening.

The growth of radio was directed by the transfer of radio knowledge through the medium itself and through newspapers. Newspapers actively, though inadvertently, built and maintained the audience needed to sustain radio, especially in its early years. In *Syntony and Spark: The Origins of Radio,* Aitken explores the balance between science, technology and the economy to conclude that "To understand the creative processes of change... our attention must focus on the ways in which knowledge is transferred" (1985, p. 335). In the case of early Canadian radio, this transmission transpired in a three-way relationship. The audience, at the core of the relationship, acted simultaneously as listener and reader to experience radio to its fullest. Dealing with the challenges of a new technology and establishing norms for listening became the task of the burgeoning audience.

Radio columns in newspapers provide testimony to the participatory involvement and enthusiasm of early radio listeners. Newspapers across the country, such as *The Globe* and *The Halifax Herald*, documented the reception conditions of the previous evening, often provided tips for assembly of crystal sets, information on what tubes were best, where to purchase parts, and how to guarantee better reception in cities such as Toronto and Halifax. The merits of local and network programmes were also extolled to listeners. This became a major consideration as interference and distance became greater factors.

Instruction and information to expand
the imagined space of radio

The imagined space of early radio is vividly revealed in the pages of the newspapers. Radio's reliance on newspapers to educate and attract its audiences is surprising because the two would seem to be natural foes. Radio posed an immediate threat to newspapers as a new means to convey news and entertainment. Immediately south of Canada, in the United States, newspapers and radio were locked in a struggle over the news until 1935, when the issue of radio news piracy reached the Supreme Court and it ruled in favour of AP in *AP vs. KVOS* (Jackaway, 1994). Newspapers were a significant proportion of the licence holders (38 per cent) when commercial radio licences were first issued in Canada, in 1922, evidence of their effort to exert control over radio (*Wireless and Aviation News*, 1922, p. 23). However, aside from a few large markets in Canada, radio soon proved too expensive to retain as merely a side operation, and the licences to broadcast held by newspapers had dropped to 13 per cent ten years later in 1932 (*Wireless and Aviation News*, 1922, p. 23). Still, the benefits of including radio content in the pages of the newspapers became evident as the newspapers with radio listings were also able to increase revenue through radio advertisements.

Radio and the newspapers did not always coexist peacefully; it took a while, in some locations, for the co-dependent relationship to develop. While some local newspapers studiously avoided inclusion or even mention of their perceived competitor, others integrated a variety of radio columns, advertisements and listings into their daily pages and often became early owners and operators of radio stations across Canada. When the *Manitoba Free Press* became the *Winnipeg Free Press* on January 1, 1931, for instance, all mention of radio ceased for a year. There were no more programme listings and no radio columns, only display advertising purchased by Winnipeg station CKY, which appeared bi-weekly on the front page. Erratic changes were a part of the early radio listings until a set format was firmly entrenched. Throughout the thirties, a constant negotiation of space occurred within the pages of the newspapers to determine where information about radio would be placed, and about the quantity and type of information that would be conveyed.

The pages of the newspapers almost immediately became a link between radio and its audience when the members of the audience were not listening, perhaps even when they were. Radio occupied space in Canadian newspapers everyday during the 1930s, but not just to provide programme listings. Radio was a fascination, a topic of consuming and ongoing interest. To quote a Halifax radio columnist, "Radio, like the automobile, has long since passed the stage of being classed as just a luxury" (Shatford, 1930, p. 14). But not every Canadian home was as well equipped as Shatford suggested; radios advertised in the pages of the newspapers were regularly priced at more than half the average man's salary. Electrification, radio reception and other factors affected the decision to purchase a radio, as well. The newspaper advertisements, though, did allow Canadians to dream about radio. Victor Radio's advertisement for the Victor "Globe Trotter" radio reads "It's a small world". This feeling of connectedness with the wider world is reported frequently in interviews with radio listeners of the 1930s. Until 1931 more than half of Canadians lived in communities smaller than 30,000 residents, so the link to the world that radio provided was very appealing, particularly in rural areas where the link to a physical community was weak. A listener in Toronto, Ontario reported that:

> Aside from the newspaper, it was the only contact we had and I remember… we all just clustered around and everybody came and listened to the radio, I would think because it was an immediate thing much more so than the papers is because you had to wait for the papers to come out. (P. Wysong, interview, October 23, 2009)

Another listener from Montreal explained:

I remember when my father first got radio, he had it all set up to listen to King George V's speech with earphones glued onto his head and my grandmother came downstairs to listen to it. (R. Leonard, interview, January 28, 2011)

The wonder of radio emerges in interviews with listeners of the 1930s, even seventy years later, particularly the idea of connecting with people within Canada, in the Commonwealth or across the world. Listener Margaret Strachan from Entwistle, Alberta, recalled she was impressed with the capabilities of radio:

I think it eventually brought people closer together because there was, you got to know a little bit more about what was going on in surrounding areas and different cities and things that you never even dreamed of, you know, if you wouldn't have radio, you'd never hear this stuff. (M. Strachan, interview, August 11, 2010)

Another listener says:

I mean, to have this in the house was like having all this new stuff people are finding out now, it was on the par with that. It was a marvelous thing to have a radio. People would come and listen to things from England and programs from all sorts of places. It opened the world. (K. Bonner, interview, December 1, 2009)

Jean Freeman, who lived on a farm in Weyburn, Saskatchewan, lived in such an isolated area that she is still struck by the connection it provided to the world:

I do remember... being impressed by the fact it was our King and he was speaking all the way from London, wherever that was, and that it was very far away and this was amazing that we could pick this up in rural Saskatchewan. It gave us a place in the world for farm families in Canada. (J. Freeman, interview, February 8, 2011)

The desire to know about radio and to hear programmes often came long before radio ownership. As Robbie Cameron of Inverness, Nova Scotia, explained, she and her sister were invited across the street to take turns listening to a neighbour's crystal set before they owned a radio (R. Cameron, interview, May 21, 2009). Pat Coulliard, who lived in Montreal in the 1930s, listened to a neighbour's radio through the wall (P. Coulliard, interview, June 22, 2010).

Helpful do-it-yourself advice on how to assemble your own crystal set radio receiver appeared in newspapers across Canada, which undoubtedly aided the growth of radio in small ways. Many rural areas in Canada were still not electrified in the 1930s, so battery operated radios became the only option in many homes. Kay Bonner of Woodbine, Ontario, remembered that their radio had "a big battery underneath it, the same kind as... a car and when the battery ran down a man in a truck came down and you would exchange it for a new one" (K. Bonner, interview, December 1, 2009). Another listener who lived north of Regina in Saskatchewan explained that her family listened to radio during the winter when they didn't need the tractor any longer and the battery could then be attached to the radio.

The acquisition of a radio or the energy to power it became a feat of ingenuity in some cases. Meanwhile parts, batteries, layaway plans and all kinds of solutions to the expense of acquiring a radio during the 1930s were offered by display advertisements, classified advertisements and radio columns in Canadian newspapers.

Attitudes toward radio varied in the columns across the country, but in general they shared a sense of the immediacy of radio and its impact on the local audience. Reception quality, boosterism and a mostly unintentional shaping of public opinion figures heavily in the content of many of the radio columns. Cities such as Montreal and Toronto had long running columns that, in large measure, provided details about the day's programmes, its stars, and local or unusual items of note. Many powerful stations were clustered around these cities and the populations had access to even more powerful ones at night. Due to the greater range of radio signals in the evenings, reception was not a problem, so local programmes were outnumbered by national and American network broadcasts in large cities.

Growing accustomed to radio

Gradually, over the thirties, listeners no longer needed to be introduced to or convinced of the benefits of radio. Radio columns were replaced with programme schedule grids and some limited programme information; in large cities these were often concerned with American network shows and stars. Smaller Canadian towns and cities experienced a similar shift until 1939, when 84.2 per cent of the country could finally enjoy regular radio reception. However, programming retained a largely local character due to limited or unpredictable reception from distant stations (Canadian Broadcasting Corporation, 1939, p. 11). The information printed in the pages of the local newspapers in smaller Canadian cities and towns maintained the flavour of the immediate geographic region of the station. Des Corry recalled that his father kept track of all the broadcasts of their favourite Vancouver boxing champion, Jimmy McLarnin, in the newspaper and would call his young son over to the radio to listen (W. D. Corry, interview, February 17, 2010). The newspaper columns that accompanied the radio listings informed and educated listeners about the local events, performers and sports. The link between audience, radio and newspapers was constantly reinforced.

By the end of the1930s, just a few major radio stations were still operated by newspapers; CHLP in Montreal was operated by *La Patrie*, CKAC also in Montreal was owned by *La Presse*, and CHNS in Halifax by *The Halifax Herald*. Halifax provides an interesting case for the examination of "reading radio" in the newspaper, due to its size, location and the common ownership of the newspaper and the city's only major radio station. The city's geographic location on the Atlantic Ocean and the eastern coast of Nova Scotia made it far removed from the broadcasting of other major cities in Canada and the United States. Partial schedules of radio stations from cities and towns smaller than Halifax were included in *The Halifax Herald* listings, but these stations did not pose a serious threat to CHNS as the only full-time Halifax broadcaster. Buffered by distance, the Halifax station chose its own course in the daytime, while the evening hours, when transmission had ceased for the day, it welcomed the broadcasts of other Canadian and American cities.

The fact that there was only one full-time radio station operating in Halifax did not present a barrier to the growth of household radio ownership in the 1930s. The mix of daily local broadcasts and American evening programming from other stations provided a sufficient volume of programming to foster that growth. In 1931, 42.18 per cent of Halifax households owned radios, falling slightly under the national average of 47.69 per cent (Canada. Dominion Bureau of Statistics, 1931, p. 980). But ten years later, in 1941, the rate of radio ownership per household had grown to 86.8 per cent in urban areas in Nova Scotia (Canada. Dominion Bureau of Statistics, 1941, p. 421). This high rate of growth was also reflected by the number of radio licenses issued in Nova Scotia, which rose from 18,027 in 1930 to 55,796 in 1940[1] (Canada Dominion Bureau of Statistics, 1934; Canada Dominion Bureau of Statistics, 1942, p. 652).

---

[1] Licenses were issued per radio, but in this period many families owned only one radio.

Halifax's ownership rate also reflects the fact that electricity was readily available. A radio station within the city limits meant that reception was far more reliable than in the furthest northern tip of the province in Cape Breton Island, where listeners report that they did not own a radio often until late in the decade[2].

The relationship between radio and newspaper in the case of *The Halifax Herald* and CHNS was a reciprocal one. The radio station was announced as the *Halifax Herald* station and CHNS placed season's greetings in the newspaper. Competing radio stations such as the island station CHCY in Prince Edward Island wished its listeners a happy new year in the *Herald* too, but CHNS read from the newspaper's comic strips over the air and shared some of the news stories. The two were aligned in the realm of opinion and content.

To inculcate, commemorate, celebrate and decry

What emerges from the pages of the newspaper is an instructive sense of what radio should be and what is it at its best. Indeed, the role of radio as an instrument to educate, entertain and inform was clear in most radio columns and government reviews of radio during the 1930s. Many of the columns, however, went beyond education, entertainment and information to inculcate, commemorate, celebrate and even express indignation.

Public commemoration assumes different forms, traditionally that of a monument. In his examination of the postmodern commemorative process, Vance argues that "The postmodern monument ... hides its politics" while the "traditional monument wore its politics openly, for all to see - it made no bones about the agenda it was seeking to push" (Vance, 2005, p. 195). In the same spirit as the traditional monument, radio columnists remarked on what they perceived as radio's momentous progress. Evidence of commemoration of Canadian radio's foundational changes was obvious and celebratory within the pages of its newspapers. The unhurried pace of decisions about regulatory frameworks in Canada guaranteed that many of

---

[2] The Aird Report published 1929 first recommended the establishment of a national network in Canada. A variety of delays meant that it was not until 1939 that the federal government extended the CBC radio coverage across the country in time for the Royal tour, making the purchase of a radio more attractive to those furthest from main centres. In particular the completion of major regional stations in Watrous Saskatchewan and Sackville, New Brunswick in 1939 accounted for most the change.

the "firsts" in Canadian radio history came late, but were duly noted in the pages of the newspapers, especially those with radio columns. While many countries around the world had already settled on public or private broadcasting and network systems by the 1930s, Canada was still in the throes of indecision. A Royal Commission in 1928 offered recommendations, and Special Committee National Commissions investigated radio in 1932, 1934, 1936, 1938 and 1939. The discussion of the ongoing debate about radio added significantly to radio columnists' ability to comment on Canadian radio history and expressed a sense of everyone's role in the process.

Commemoration of connection

The connection between Canadians listening to the radio on a national, regional and local level was frequently noted. During the early 1930s independent radio dominated, although the formation of small regional chains was beginning, along with the entry of American network affiliation in Canada and the railway network. Shatford notes in his *Listening In* column in *The Halifax Herald*:

> … it may not be generally known that the Canadian National was the first railroad in the world to adopt radio reception as a regular service on its trains. Sir Ernest Rutherford was the first to succeed in transmitting wireless signals to a moving train, carrying out his experiment on the Grand Trunk between Montreal and Toronto as far back as 1903. (Shatford, 1930, p. 12)

In his commemoration of radio's time and place, Shatford goes on to explain that twenty years later the first radio-telephone message was broadcast to a moving train so that W.B. Robb, vice-president of the Canadian National Railways could be broadcast. He also explains that 72 cars were permanently wired for radio across the country. These kinds of networks were especially interesting to regions of Canada that could not receive network programmes from the United States or Canada on a regular basis.

American network affiliate stations were established in both Montreal and Toronto by 1931 providing access to many North American broadcasting events and programs. However in the rest of Canada, commemoration of radio events was frequent in radio's early stages, sometimes, simply because the use of telephone hook-ups across the country to create a network was so difficult. Shatford also exclaimed:

> A new page in the history of international relations was written Christmas day when listeners throughout North American heard musical expression of good will from Holland, Germany and England. (Shatford, 1930, p. 12)

The columnist goes on to list the musical contributions from Europe and the United States and concluded that:

> The interchange of programs added laurels to the brows of engineers who for the past half dozen years have been at work in making come true the dream of sharing musical programs with the world. The quality of the programs surpassed anything previously heard in this country from across the watery waters of the Atlantic. Furthermore, America's carefully prepared programs were heard clearly abroad, according to cables received shortly after the program. (Shatford, 1930, p. 12)

This radio column was not only commemorative, but celebratory in tone. *Listening In* regularly conveyed a sense of "great steps in radio history" in the pages of *The Halifax Herald* during the 1930s, as broadcasting passed through various stages.

When the Canadian Radio Broadcasting Commission (CRBC) started broadcasting across the country in 1933, it received largely critical press but as news copy rather than in the radio columns. By that time the technical marvels of radio were largely taken for granted by newspapers, if not by first-time listeners. The broadcasting of both English and French programmes on the same network became a point of contention, especially as reported in western provinces that did not appreciate their programming time being taken up by French-language programming which only a minority of the population could understand. The mixed reviews of this fledgling national network served to influence public opinion, and these critiques are considered to have been a major factor in its quick demise.

The anticipated arrival of the Canadian Broadcasting Corporation (CBC) became another moment in Canadian radio history commemorated by radio columnists. Just prior to the start of CBC broadcasts in 1936, the radio columnist in the *Winnipeg Free Press* inundated readers with blow-by-blow coverage of the network's plans. Winnipeg was long considered the gateway to the West and, as such, became the location chosen by the CBC to audition the musical talent that would obtain national programmes. Daily reports of the bands that travelled to Winnipeg from all over the West described in detail their performances, and provided background information, tidbits from the CBC and, finally, ratings of their music and suitability to the audience. A great sense of pride exuded from the columns, along with approval of the national network's attempt to integrate the regional identity of the West into this new national network. Canadian national, local and regional identities were inculcated continuously by a variety of columns across the country, and the branding of CBC as a national network that would reflect all the regional identities gained repeated approval.

In addition to the remembrance of moments in radio's regulatory and technological history, some of the pivotal events in the news that figured heavily in radio and newspaper coverage also figure heavily in the memories of those who listened. Daniel Dayan and Elihu Katz note the importance of commemoration of events such as the abdication of Edward VIII in their larger discussion of the promotion of unity and collective memory (Dayan & Katz, 1992; Couldry, 2003). A listener from Winnipeg, Manitoba still remembers when Edward VIII abdicated in 1936:

> … it came on in Winnipeg in the middle of the night… about 2 a.m. or so broadcast from London England…the Prince made his heart-breaking statement about abdicating giving up the throne for the woman he loved… who was an American divorcée and of course it was a big kafuffle because the royalty at that time were supposed to be pure and simple and all that. [When] that came on in Manitoba, my friend and I stayed up to listen to it that. That was one event of international importance that I remember on the early, early radio because that's where you got the news and lots of publications about the fact that this was going to be aired so we stayed up hear it. (R. Hughes, interview, October 20, 2010)

The layers of new technology and historic events affected the quotidian lives of listeners, but also affected their larger sense of identity as part of the former British Empire, and as members of the Commonwealth. Simon Potter argues that during the interwar years "relationships between Commonwealth [radio] broadcasters reveal the conditions that facilitated, but also set the limits to, the forging of closer cultural ties" (Potter, 2006, p. 425). Whether Commonwealth relationships were strengthened or weakened, the stories of monarchy, Christmas wishes from the King and other connections to the sense of their larger identity were featured heavily on the early radio stations (Vipond, 2003; 2010). The connection to England was very much alive for listeners who had only recently arrived in Canada, as noted by Kay Bonner from Woodbine, Ontario:

> My parents came from England... so for them, news from there for
> instance, when old King George V was dying, that was broadcast all over
> the world, hearing the King's life was passing peacefully to a close...
> [T]hen of course he died and we could listen to the funeral service...
> then... [the] Prince of Wales... was going to become Edward VIII, but he
> abdicated and that was another time when everyone listened in on the
> radio. (K. Bonner, interview, December 1, 2009)

Robert Cupido (1998) argues that the Diamond Jubilee of confederation
served to encourage the growth of national feeling as much as the yearly
greetings from the King at Christmas. But international news events such
as the abdication continue to survive in the memories of the listeners
interviewed, as do national crises such as the Moose River Mine disaster
in 1936 (Webb, 1996).

The King's Christmas message became a tradition that necessitated
hook-ups across Canada. Mary Campbell lived in Winnipeg, Manitoba and
remembers it well:

> The Christmas Day's King's Speech... with the reports from the empire...
> We listened carefully to that. It was unusual to hear voices from around the
> world and they ran a whole program so you'd hear someone vaguely
> speaking from Australia or South Africa or India leading up to the King's
> speech. It was very special programming for Christmas day. You didn't get
> to open presents until after you listened to the King's speech. (M.
> Campbell, interview, January 18 2011)

Finally, indignation about certain developments in radio figured in the
regular columns of the newspaper and was echoed by the listeners. In
1930, the radio column in *The Halifax Herald* reported in a highly
indignant way that the annual good wishes of the King at new year had to
be broadcast across Canada through an American national network rather
than over a Canadian network. The sense of Canadian national identity and
structure that figures so heavily in the policy documents of the time was
reinforced by the newspapers and radio.

This situation is echoed in the case of the broadcast of King George's
death six years later, as told by a listener in Coaldale, Alberta:

> When King George died, I happened to be driving around my car at three
> in the morning and they don't know why but I guess I have been to a party
> or something and heard that King George died on [an] American station

and I think it was New York but I'm not sure but one of those fluke things you get every so often, stated that George V just died. So I rushed to where the main newsman was in the city and knocked on his door at three o'clock in the morning and woke him up. He was kind of mad about that. And I said I'm here to help you, George V has just died. He hurried and dressed and I waited and drove him back to see if the local station reported it and he checks that it was true and then started to prepare to report and it was the only radio station in the area that went on the air with this big news. (S. Reed, interview, January 17, 2011)

This listener's recollection provides valuable insight into the strong sense of connectedness to news on the world stage that audiences came to expect from the radio. Though Coaldale is located in a rural area approximately one hundred kilometres from the American border to the south, the expectation was that a Canadian radio station should carry news about events in the Commonwealth.

Conclusion

Memories of radio listeners attest to the increasingly direct connection between the audience and radio. Initially newspapers provided guidance for time and places on the dial where listeners could tune into boxing matches, sporting events and royal wishes. Once the connection between the audience and radio was cemented, newspapers ceased to perform an active role as intermediary. Newspapers had commemorated radio's beginnings and Canada's place within the development of the medium. They inculcated a sense of the audience and community by creating a common ground for the listeners (Anderson, 2006). However, the newspaper was foremost in its solidification of the connection between the radio listeners and their larger world, whether it was through a connection to a locality, region, Canada or the Commonwealth.

By the mid-thirties the combined effects of radio broadcasting, newspaper reports and columns about radio had produced a listening public with expectations that radio would reflect their sense of place in the world. It appears that 1936 was pivotal for Canadian radio. It marked the beginning of the CBC, the death of King George V, the abdication of Edward VIII and changes in the relationship between radio, the newspaper and the audience. The audience had reached a point at which information about radio assembly was no longer essential, and radio personalities were well known. Celebrations marked by radio programming had become a tradition, and the audience expected reports of world events to which they felt a connection. Newspapers ceased to provide the same details about the reception conditions and about radio performers. Schedules became more predictable and many were published as a grid, dependent on routine, and only highlights were the subject of a column or featured in supplementary description. The reciprocal relationship between radio and newspapers generated an independent audience with a clear sense of what radio should be.

References

Aitken, H. G. J. (1985). *Syntony and spark: The origins of radio.* Princeton, N.J.: Princeton University Press.

Anderson, B. (2006). *Imagined communities: Reflections on the origin and spread of nationalism.* New York: Verso.

Canada Dominion Bureau of Statistics. (1931). *Seventh census of Canada.* Ottawa: The King's Printer.

—. (1934). *Bulletin: The radio industry in Canada, 1931.* Ottawa: King's Printer.

—. (1941). *Eighth census of Canada.* Ottawa: King's Printer.

—. Department of Trade and Commerce Canada. (1942). *Canada year book 1942.* Ottawa: King's Printer.

Canadian Radio Broadcasting Corporation. (1939). *Canadian broadcasting: An account of stewardship.* Excerpt from Canada Parliament House of Commons Special Committee on Radio Broadcasting. *Minutes of Proceedings and Evidence, Nos. 2 and 3.* Ottawa: King's Printer.

Couldry, N. (2003). *Media rituals: A critical approach.* London: Routledge.

Cupido, R. (1998). Appropriating the past: Pageants, politics, and the Diamond Jubilee of Confederation. *Journal of the Canadian Historical Association, 9,* 155-186.

Dayan, D. & Katz, E. (1992). *Media events: The live broadcasting of history.* Cambridge, MA: Harvard University Press.

Jackaway, G. (1994). America's press radio war of the 1930s: A case study in battles between old and new media. *Historical Journal of Film, Radio and Television, 14,* 199-314.

Potter, S. (2006). The BBC, the CBC, and the 1939 Royal Tour of Canada. *Cultural and Social History, 3,* 424-444.

Shatford, L. L. (3 January 1930). Listening In. *The Halifax Herald, 55,* p. 12.

—. (8 January 1930). Listening In. *The Halifax Herald, 55,* p. 14.

Vance, J. F. (2005). Documents in stone and bronze: Monuments and memorials as historical sources. In J. Keshen & S. Perrier (Eds), *Building new bridges / Bâtir de nouveaux ponts: Sources, methods, and interdisciplinarity / sources, méthodes et interdisciplinarité* (185-195). Ottawa: University of Ottawa Press.

Vipond, M. (2003). The mass media in Canada: The Empire Day broadcast of 1939. *Journal of the Canadian Historical Association, 14,* 1-21.

—. (2010). The royal tour of 1939 as a media event. *Canadian Journal of Communication, 35,* 149-172.

Webb, J. A. (1996). Canada's Moose River mine disaster (1936): Radio-newspaper competition in the business of news. *Historical Journal of Film, Radio and Television, 16,* 365-376.

Wireless and Aviation News, (1922 April), p. 23 as cited in Vipond, M. (1992). *Listening in: The first decade of Canadian broadcasting, 1922-1932.* Montreal-Kingston: McGill-Queen's University Press), p. 21.

CHAPTER THREE

NOT OVER HERE! HOW BRITISH ELITES USED NATIONAL NEWSPAPERS TO ENGAGE IN DEBATES OVER THE INTRODUCTION OF LICENSED COMMERCIAL RADIO STATIONS

RICHARD RUDIN

Introduction

The public debate on whether Britain should have licensed commercial radio stations, which lasted from the end of the 1950s through to the mid-1980s (when the Labour Party abandoned its long-held opposition to such services), is not just of interest in terms of media history and public policy on broadcasting but was, as this chapter shows, an arena in which deep philosophical, political and cultural issues and policies were debated. Just as the superpowers used "proxy wars" to test each other's resolve in the Cold War, as Johns (2010) argues, the debate about commercial radio can best be understood as shadow boxing for much more fundamental changes in society, culture, economics and politics. The liberal-left may well have been vindicated in seeing commercial stations as leading the charge for a switch from the post-war consensus of a culture of social democracy (and even socialism), to free-markets and individualism, supported by a market economy. What is striking when looking at how the issue was debated and reported is how much language is contested in ideological terms: what is "freedom", what is meant by "independent"? And, in a more conspiratorial vein, what and whom are really behind the push for commercial radio and, on the other side, what are "they" (the establishment, old and new) so afraid of by the prospect of seemingly harmless, if bland, commercial fare?

A feature in the debate about commercial radio was a sometimes explicit, but often implicit, anti-Americanism or, at the very least, a firm belief in the superiority of British broadcasting compared with its manifestations across the Atlantic. The pre-World War II commercial stations such as Radio Luxembourg, broadcasting in English from the near continent of Europe, had achieved very substantial audiences, especially on Sundays when the BBC broadcast mostly "serious" fare, and are fondly remembered. In addition, many British servicemen and civilians had sampled commercial radio fare via the American Forces Network (AFN) during the war. Indeed, the BBC's monopoly was technically broken in war-time, as the service – despite much BBC opposition and lobbying – was allowed to transmit within the UK in the lead-up to D-Day in 1944. The UK's own Forces Programme transmogrified post-war into the Light Programme and drew on some of the techniques and programme styles of the AFN, albeit in a much watered-down fashion. The British elites, though, were more likely to have visited the US and could therefore draw on first-hand experience or, at the very least, have been in discussions with those who had. By the mid-1960s the enormous success of the offshore "pirate" radio stations, which were to have continued appeal and be regarded with great affection by much of the public decades later (Rudin, 2008), provided a new challenge to the BBC's monopoly and exposure to US inspired, and often financed, radio services.

This chapter shows how leader columns, letters pages and features in the biggest circulation daily newspaper of the day and several of the leading so-called "quality" broadsheets contributed to these debates in the public sphere in the lead-up to the first of the licensed commercial stations. It will also explore the difference in anxieties held by the tabloid and the broadsheet papers as to the impact commercial radio would have on existing stations, and what should be the form and content of the new services. It is clear that the newspapers had vested interests in helping to scupper plans for new stations, which would be competitors for advertising revenues. This chapter shows that, in the case of the "quality" press, such self-interest could be disguised with more lofty concerns about the 'vulgarisation' of British culture. The chapter begins with even more conflicted views from then biggest-selling and left-leaning daily paper, with a largely working-class readership, which was caught between a desire to attack the 'stuffy 'and 'pompous' BBC and concerns about the invasion of free-market values into UK broadcasting.

Support for commercial radio from the *Daily Mirror*

As post-war austerity began to ease, the intriguing possibility of advertising-funded or sponsored radio began to permeate the public sphere. In part, this was a conscious reaction to the fact that, while the rest of society changed rapidly under the post-war Labour government, the BBC was characterised as unreformed, overly paternalistic, elitist and morally high-toned for the new supposedly meritocratic age. The biggest selling newspaper for the time, and one that had clearly promoted itself as being on the side of the common man, was the *Daily Mirror*, which enthusiastically supported the post-war Labour government. However, alongside this, and to some extent in tension with it, was support for "lowbrow" entertainment and the feeling that the workers, as well as having "free" health care, welfare benefits and security from the cradle to the grave etc., were also entitled to diversion and entertainment. "Sponsored" radio, with its emphasis on entertainment, could therefore be seen as a legitimate—even necessary—adjunct to the respected but worthy BBC. The newspaper's proprietor, the aristocratic Cecil King, was a keen supporter of "sponsored broadcasting" and a major influence in the Labour Party until his sacking by the board of the paper's controlling company in 1968 for his notorious self-penned front-page leader ("Enough is Enough", 1968) calling for the resignation of the Prime Minister Harold Wilson and the Labour government's replacement by a National government of indeterminate nature and composition (but most likely containing a certain Cecil King). It is surely no coincidence that the *Mirror*'s enthusiasm for commercial radio demonstrably collapsed after this dramatic *coup*. By the early 1950s, though, the paper had started actively campaigning for commercial services, despite the fact that the Labour Party consistently and vociferously opposed commercial radio and television right into the 1980s. This tension between socialist abhorrence at the radio waves being put in the hands of capitalists, and a populist, even hedonistic culture, was consciously and openly worked out within its own pages, when the newspaper's editorial line was placed in opposition to that of its most famous columnist "Cassandra". In April 1952 ("Mealy-Mouthed Arrogance", 1952) it chose the topic as its sole front-page editorial lead and asked rhetorically:

> Will the country go to the devil if the people are given a choice of radio instead of being bound to the B.B.C.? Those who oppose gingering up the B.B.C. by sponsored competition think it will. They are in full cry against giving the public an independent alternative. Their attitude is roughly that the B.B.C. knows best ... and the public must be prevented from indulging

its low and tasteless inclination to roll in the sewers.... We have enough trust in democracy to believe that a people who have the right and the capability to choose their Government, their newspapers, their books, their films and their plays should not be insulted by being told they need governesses to select their radio.

This editorial is interesting, not least because of the use of the word "independent" – to be taken up by Conservative governments to describe both commercial radio and television. The penultimate paragraph also somewhat eerily predicted that the ultimate victory for the cause of commercial radio would result in a very much toned-down version of its US counterpart:

We believe that competition would give the public better radio and the existence of the B.B.C., plus public taste, would prevent the abuses of American sponsored radio from being tolerated here.

The final paragraph pointed the reader to the "Cassandra" column on page six of the same edition ("Ariel – Don't Go Back!", 1952) for the opposing point of view[1], worth quoting at length because it acknowledged the inadequacies of the BBC, as well as the Corporation's superior tone which must have grated so much on so many *Mirror* readers, while at the same time directly citing the US experience as the clincher argument against commercial radio. It is also a column that does not patronise its audience; it assumes that they will be familiar with the legend of Prospero and Ariel, allusions to whom begin and end the column.

If yesterday morning outside Broadcasting House, in Portland Place, you observed a moody figure staring with rapt and intense expression at the stone figure of Ariel over the B.B.C.'s main door, it was this correspondent at his devotions. My supplications went like this:

"Dear Ariel, stop 'em! Guard the skies and sweep 'em away with all their basement-bargains, their cut-price offers and the raucous clatter of their hawking…"

I have recently heard and seen commercial sound broadcasting and television on the East and West coasts of the United States. I pronounce it abominable… The commercial sound broadcasting is, for 90 per cent of the time, intolerable. The rest of the time it is merely banal, distasteful and wearisome. Day and night the hideous yelling and selling goes on by bawling, insincere young men and throaty-voiced girls. Even the news is

---

[1] "Cassandra" was the pseudonym of left-wing journalist William (Bill) Connor.

hedged about and wired in by some pickle manufacturer or sanctimonious Hollywood embalmer. There is no escaping it – an unbearable yelping and yapping about buying and selling that is a kind of grinning third degree from which there is no escape ... The B.B.C. with all its stuffiness, with all its bland lack of courage and with its wretched passed-to-you policies is brilliant, bold and truly delightful compared with this maelstrom of mercantile muck.

Five days later the *Mirror* triumphantly reported a victory in public support for the overall editorial stance ("Sponsored Radio – Readers reply to Viewpoint", 1952):

The "Daily Mirror" and Cassandra took opposing views this week on a subject of national controversy – should Britain have sponsored radio programmes. Readers were asked for their views. Ten out of every seventeen who wrote agreed with the "Daily Mirror" that sponsored radio is justified.

The leading letter explicitly linked the idea of commercial radio with the US and, crucially, the virtues of free enterprise in broadcasting and in the wider society:

In all towns in America there is always a choice of programmes, gay and grave. Competition means enterprise and value... Competition certainly leads to enterprise. When I reached Salt Lake City where I was lecturing, a mike was rushed down to the platform to meet me, and gave a greeting to the State red-hot ... impossible in England. No enterprise – because no competition.

The pre-war continental stations had amply demonstrated that commercial radio could serve up enjoyable dollops of lighter fare but this same correspondent spoke warmly of American stations relaying famous operas, supported by unobtrusive sponsorship announcements at the beginning and the end. The view that commercial broadcasting need not necessarily or always equate with low-brow, trivial fare was further supported by another correspondent:

Tell Cassandra to keep on the networks of Boston University, Columbia, and the Voice of America and learn something. American broadcasting has a kick in it. The B.B.C. could do with a few kicks.

If kicking was to be done, then the BBC—self-regarded as the purveyor of an excellent, impartial and comprehensive news service—was further attacked by a correspondent who found its journalism distinctly unimpressive:

> Most B.B.C. programmes are tripe. All we hear is: "The programme you
> have just heard was recorded." Even the news, which is repeated over and
> over again, can be read in the morning's *Daily Mirror*. It is only recently
> that I have discovered Radio Luxembourg, and I was astonished at its
> friendly, homely atmosphere. If this is an example of sponsored radio, then
> I am all for it. We have free speech. Let us have free listening, too.

The hand of the paper's Chairman and Managing Director in this
positive approach to commercial radio can be shown by his comments at
the following year's AGM. *The Guardian*'s report ("Company Meetings –
The Daily Mirror Newspaper", 1953) of the meeting directly quoted Cecil
King as arguing:

> In the public discussions on the subject it is often stated that the British public
> would not stand for commercials in its radio entertainment. This, I maintain, is
> nonsense.

The tensions between the views of a Lord who clearly favoured free
enterprise, and his paper - which supported trades unions - are evident in
King's assessment for the reasons the Labour Party opposed commercial
radio:

> Their opinion seems a mixture of Socialist doctrine, Puritanism and a fear
> of the political bias of commercial programmes. It is natural that a Socialist
> Party would dislike seeing private enterprise re-entering any sphere that
> has been nationalised; but this argument cannot be considered very
> weighty. The puritanical outlook is presumably at bottom opposed to all
> entertainment and prefers the smug qualities of the B.B.C. to the livelier
> indiscretions of the commercial world. It is likely that many of the radio
> advertisers would be anti-Labour, but doubtful if they would let their
> political bias seep into their programmes and still less likely that any such
> bias would prove effective.

The fear that advertisers or sponsors would directly interfere in
editorial content was to be allayed by the strict separation of the two areas
in both of the Acts which introduced commercial television and—much
later—radio into the UK and limited commercials to 'spot' advertising in
programmes' "natural breaks". Thus, one of the main objections to
commercial broadcasting had been removed in the UK, but these strict
regulations did not seem to pacify many critics.

By 1959 the campaign for the breaking of the BBC monopoly in radio, as it had been broken by "independent" television four years earlier, was gathering momentum. In March 1962, as the final touches to the Pilkington Report on the future of broadcasting were being made, the BBC's Director-General, Hugh Carleton Greene, spoke at an awards' dinner in Washington, DC. He was scathing about the output of US radio and argued that American values of "freedom, democracy and competition ... were strikingly misused and were intended to disarm the critical faculties" ("BBC Chief Reviews U.S. Radio", 1962). Furthermore, reported *The Times'* special correspondent, he criticised those who thought that any interference from government would enslave the broadcasters but who seemed content that the commercial sector was "in bondage" to economic interests. He directly challenged the defence that such services were simply giving people what they wanted:

> This, with "democracy" and "trusting the people", was part of the simple faith preached by men who were not at all simple – that what most people want all people should have. It was an abuse of language; they were in fact concerned with tyranny, the tyranny of the ratings ... there was no real triumph in merely giving some mild pleasure or soporific to people too indifferent to switch the programme off.

Thus, Greene, although generally regarded as one of the most liberal DG's in the Corporation's history, explicitly aligned himself to the Reithian values and ethos seemed to be under threat now that the principle of a monopoly had been broken in television and there might seem to be little logic in preventing competition in radio.

In the event, though, Pilkington backed the BBC, both in television (the third channel was one of the other main issues) and in local radio, so the battleground moved to the *de facto* introduction of commercial radio through the unlicensed, offshore "pirate" radio stations of the mid-1960s. The success of these stations led to a number of broadcasting policies and alternatives being discussed by the Labour government, which came to power in October 1964, just over six months after the start of *Radio Caroline*. There was a clear wish in the new government to close the stations down, but there were legal, technical and political problems (annoying so many voters) to consider.

How the British intelligentsia voiced its opposition to commercial radio

One of the chief ways in which the public sphere was engaged in the issue of commercial radio was through the letters' pages of the "quality" press. This section of the chapter will concentrate on the contributions made by two leading intellectuals who were then both at the University of Birmingham, forming the Centre for Contemporary Cultural Studies (CCCS), or the so-called "Birmingham School" of sociology.

Richard Hoggart and Stuart Hall first became engaged with the issue in 1964, within the first six weeks or so at the start of broadcasting of *Radio Caroline*. The two wrote to *The Guardian* (Hoggart & Hall, 1964, May 18) to protest against what they called the "serious problems" caused by such broadcasting ships. The chief concern of the writers was that commercial radio, which they said for "substantial reasons", had not so far been licensed in Britain, would become legalised and normalised through inertia by the government and the lack of critical engagement in the issue by the "intellectual press" against such services. The pressure to license the stations, they claimed, arose simply from "financially interested parties".

Less than two weeks later the pair developed their arguments against commercial radio in a substantial letter published in *The Observer* (Hoggart & Hall, 1964, May 31). In this, they were keen to delineate between the arguments for commercial radio and local radio - the latter of which they approved:

> Local radio—democratically established—could provide a useful service of local news and views, sports coverage, education and so on; and such stations could at other times provide light or classical music or whatever else seems likely to interest listeners. The crucial point is that these stations, each under the control of the station manager whose brief was, *and was only*, to serve his town or area as fully and flexibly as possible, could have a wide range of materials and of levels in output (original emphasis).

The sociologists countered the claims being made by some groups proposing commercial stations that such desirable additions to local communities could be provided by advertising-funded services. In an argument that was to be repeated and refined many times over the following decade or so before such stations were established in Britain, they argued that:

> The logic of their commercial position—all precedents show—would
> make them progressively forego range and variety of service in favour of
> concentration, aimed at reaching as much of the time as possible the most
> tightly packed group of likely consumers. The obvious example is Radio
> Luxemburg's concentration on disc jockey programmes with a teenage
> audience - a solid prosperous body of consumers for cosmetics, chocolate,
> records. How much interest would the pressure of his balance sheet allow a
> commercial station manager to show in, say, the not so prosperous elderly,
> or in the "serious-minded" who want occasional educational programmes
> rather than endlessly careless glamour, or in many [sic] other group which
> is unattractive to advertisers?

The argument that such stations merely provided what the public
wanted, demonstrated, they said, "a miserably bankrupt view of social and
individual responsibilities". The conclusion of the letter bears particular
examination, as it is perhaps the best exposition of the arguments against
the introduction of what seems, in contemporary thinking, a fairly
innocuous proposition to have entertainment-led commercial stations in
opposition to the BBC. The writers clearly believed that a crucial moment
had arrived and that, while the paternalistic attitudes of the BBC might be
objectionable, it would be almost tragic to take what they termed the "rule
by cash registers" approach.

> It has taken centuries to forge even those democratic structures—in law, in
> politics, in the provision of education elsewhere —that we now have. Mass
> communications within prosperous centralised democracies are a new
> phenomenon and we have not yet found the organisational forms, or the
> general temper, which will allow us to use them in the best ways.

The final paragraph explicitly took on the battle over the lexicon and
historical resonances in the words "free" and "independent" that the pro-
commercial lobby had appropriated as the main moral and political case
for their argument. The "free press" —that is, the freedom to publish
without prior approval or licensing and without the taxation on the
economic and physical means of publication, and subject only to the
general law of the land—had only been established in Britain since the
1840s and after several centuries of struggle against the authorities. But,
unlike newsprint, there is a genuinely limited number of frequencies and
channels available for broadcasting and therefore government had a
legitimate reason for licensing and regulating radio services. The use of
the term "free radio" was an attempt both to indicate freedom from control

and censorship, and freedom at the point of use: a powerful twin argument for alternative and competitive services to the BBC—which was funded by a licence fee on radio, and then TV sets—and the breaking of its monopoly. Similarly, the word "independent" had been adopted to describe commercial television, carrying with it attractive and virtuous connotations of independence from the authorities and, by contrast, implying that the BBC was a tool of the state. Hoggart and Hall clearly recognised the power of these words and explicitly sought to challenge their use and to expose what they regarded as the true effects and results of commercial radio, which, if allowed would mean:

> ... we shall be committing an act against democratic growth roughly comparable to re-instituting the taxes on knowledge. *"Freedom and independence"* does not lie in giving a handful of individuals the chance to make large profits, but in finding forms of organisation which will keep this powerful means of human communication free from exploitation by any manner, whether cash or for ideology, and therefore free to meet the varied needs and potentialities of a changing society.

The following year, Hoggart was a witness for the BBC in its backing of a case involving Phonographic Performances Limited (PPL), one of the chief licensing bodies for the "performance" of recorded music and the one which policed the "needle-time"[2] agreements with radio services. Within a year of its opening, the Isle of Man's *Manx Radio* was in dispute with PPL over its "needle-time" allocation and took its case to a Performing Rights Tribunal held on the island, supported by evidence from the Market Research Society. *The Guardian* reported witness Gerald Stacey's statement that this evidence showed that the new local commercial service[3] was not just enjoyed by "pop-loving teenagers";

---

[2] "Needle-time" was the agreement between the music copyright bodies and the broadcasters in the UK which, until the 1990s, limited the amount of commercially-released recordings that could be played on a radio service or services over a given period. It resulted from a fear by the Musicians' Union, in particular, that unrestricted use of records on radio would result in reduced employment of 'live' musicians.

[3] The Isle of Man, situated off the north-west coast of England, has a complicated and sometimes contentious constitutional relationship with the United Kingdom. It is proud of its ancient parliament, the Tynwald, and its autonomy over most domestic policies, which enabled it to establish its own commercial radio station in 1964, but is subservient to the Westminster parliament and government over foreign matters. This 'split' governance led to a constitutional crisis in 1967 when the Tynwald declared that the UK government's Act against the 'pirate' stations did not apply to the island, which wanted *Radio Caroline North* to continue off its

indeed he claimed "My researchers showed that housewives were the most satisfied" ("Choice of the Manx Wives", 1965). Of the 400 people interviewed in the survey, 80 per cent listened to *Manx Radio* – a third more than to BBC stations overall and almost twice as many as to *Radio Caroline*. The open and, one might argue, improper support by the BBC for a licensing body against the country's first legal commercial radio station was justified by the Corporation's witness, Michael Kempster:

> We are not involved in any conspiracy to preclude the people of this island from commercial broadcasting. We are concerned in maintaining one thing - that it was not unreasonable for the licensing body to impose the terms and conditions which it has.

Kempster's argument seemed to be that increasing the station's "needle-time" would diminish what he described as "a good and balanced programme". He clearly supported the BBC's line as it enjoined with the great debate over the next seven years over proposals to legalise commercial radio across the rest of the UK: "Commercial sound broadcasting may seek the same things [as the BBC] but they are bound to be tempted to produce programmes appealing to the greatest number of people for the whole time".

Professor Hoggart was "surprised" at the limited amount of time the station devoted to the island's affairs and, perhaps rather patronisingly, opined: "It is splendid to think the Isle of Man has got its own radio station but one has the feeling there are so many lost opportunities here".

It is hard not to conclude that the opponents of commercial radio, including the BBC and the intellectual elite, were terrified that *Manx Radio* would prove them wrong and that commercial stations could both provide good local content as well as entertainment and diversion.

Six years later Hoggart was delivering the annual Reith Lectures[4] on the BBC and although this was the period when some 20 BBC local radio stations had been established and the BBC still had a monopoly over all radio broadcasting, Hoggart was concerned, as *The Guardian* reported, that the emphasis on "the local" was to the detriment of the national

---

coast. The UK government used an Order in (Privy) Council to declare that the Act did apply to the Isle of Man.

[4] Every year the BBC invites a leading figure to give a series of broadcast talks on "significant contemporary issues" related to their field of expertise.

("Need For National Radio", 1971):

> ... one of the benefits of broadcasting is exactly that it allows a nation to speak to itself (and nation to speak to nation). It is a reversal to parochialism to think that small-scale communication is a substitute for large scale; the two kinds are complementary.

With the Conservative government's bill to introduce local commercial radio having just been published, Hoggart returned to his central fear that local radio generally was a Trojan Horse, and that national channels would be handed over to "ideological types and the commercial sharpshooters so that they can distort them for their own ends".

Leader articles in *The Times* and *The Guardian*

The leader articles in the two "quality" papers considered here had remarkably similar lines on the desirability of the BBC or other non-commercial interests being charged with developing any local radio services. The key periods of debate were in two distinct phases: reaction to the initial broadcasting by the "pirate" radio stations in 1964-67, and then in 1970-72, prompted by the surprise win by the Conservatives in the 1970 general election which led to an early White Paper and then a bill to introduce local "independent" services.

Some two months after the start of the first two offshore "pirate" services directly aimed at a UK audience, a Leader article in *The Times* ("Do The Public Want It?", June 1964) criticised the Conservative government for suggesting a review of "local sound broadcasting", apparently prompted by "the assiduous way in which the pirate ships have been courting publicity", and argued that such stations "have no relevance to local sound broadcasting". The only thing such stations had demonstrated was that there was "a voracious appetite for endless streams of pop music." Rather than licensing such services, the paper argued (contempt for popular music and its followers is obvious throughout the archives) that an "aesthetic case" could be made that there was already "far too much of it on existing radio and television channels".

Also in June 1964, *The Guardian* argued against legalising such a service. Its main leader ("Don't Let Caroline Come Ashore", 1964) asked rhetorically:

Does any community feel culturally deprived in being denied its own disc jockey? And is the commercial radio lobby really animated by a sense of cultural mission? ... If a demand for local radio ought to be satisfied, the BBC seems the body best qualified to do it. ... In its local radio trials, run on closed circuit since 1962, the BBC has emphasised public service. It would make its local station a forum for local debate; it would provide local news and information ... There might be more demand, on an audience count, for commercial candyfloss. But on likely quality the BBC has it.

By November 1966, with the Labour government now seeming serious about a bill to put the "pirate" stations off the air, *The Times* had got its ear to the ground at both the BBC and government about how they were to be replaced. A Leader ("Lighter Than The Light", 1966) acknowledged that if the government valued "the good opinions of young voters they had better make other arrangements for providing the same sort of noise". The three options seemed to be a new pop channel run by the BBC, a "pop" broadcasting authority financed by advertising on a wavelength currently used by the BBC and, thirdly, commercial local stations. The paper clearly supported the first proposal, which was indeed adopted – partly on the grounds that if local commercial stations went ahead:

... there would be no room for those programmes of local utility and interest which the advocates of local broadcasting have in mind. The civic attributes of a possibly hopeful experiment would be cancelled in advance.

The Leader acknowledged, though, that the question of "needle-time" was going to be a problem – "the main brake on any legitimate provider of the drug".

By the following month *The Times* was clearly relieved that the government White Paper had indicated that the BBC would provide both local radio and a national "pop" service and that, in the words of its Leader column ("Pop Goes the BBC", 1966) "commercial interests are warned off". However, the wider participation and funding from other public bodies to the planned BBC services were viewed with alarm:

The arrangement would also appear to introduce undeclared sponsorship of editorial matter, which is as objectionable in broadcasting as it is in journalism.

In the event, the nine experimental BBC local radio services were partly funded by local government; indeed, the location of the first services was mainly determined by which local authorities would put up the money to contribute towards the costs of the station – "radio on the rates".

The following spring, as the government's bill to quash the "pirates" was proceeding through parliament, *The Guardian* was much more enthusiastic ("Local Radio With A Purpose", 1967) about such local civic participation. Its strong historic links with Manchester no doubt inspired, or at least added to, its enthusiasm for the prospect of that city having a local station and it was dismayed that the Conservatives, who had just won power there, were baulking at the idea of local taxes paying for the service.

> The pirate stations have demonstrated that their kind of radio can be a success commercially, but their output has been mainly mindless chatter and banal music. The BBC has long since demonstrated that public service radio can provide a rich diet.

In a rare creative proposal it suggested that if the local (government) council would not provide the cash, then the city's main university might: "It could be something new and quite different". Two years after the first local stations started broadcasting, *The Guardian* in November 1969 (Shearer, 1969) argued that:

> If local radio is to be expanded the BBC is the organisation to run it. The blandishments of the commercial lobby should be resisted ... The commercial lobby gives the game away ... local radio would be of lower quality, as indeed it would be if it were run by commercial interests. These interests would almost inevitably have to appeal to the lowest common denominator of public taste to secure the maximum audience.

Just one week later, the paper ("The BBC and Its Standards", 1969) re-iterated the claim that commercial radio would mean a debasement of standards and quality, and also brought in the economic arguments:

> It would be helpful if the Conservative spokesman, Mr Paul Bryan, would abandon his absurd verbal distortions in suggesting that the public would get commercial radio 'free'. Does he really think he can persuade people that none of the £100 millions a year spent on television advertising is passed on to the consumer in higher prices in the shops? Why would it be otherwise with radio?

In a leader ("The Toothless Pirate", 1969) following a House of Commons debate on broadcasting policy, *The Times* found itself, criticising *Radio One*—in a rather unexpected way, given the paper's stated loathing of the offshore stations and 'pop' music culture—for being: "… a pop programme of quite exceptional dreariness. A toothless pirate …", but thought it was still unlikely that local radio—any local radio—was necessary and that it should only be expanded if there was enough money to do it without affecting expenditure on such things as the Corporation's orchestras. However, expanding the Corporation's local services, it concluded, would be "better than the thoroughly undesirable Tory plan for commercial local radio".

Barely three months after the 1970 General Election, *The Times* was dismissive of the arguments put forward by those advocating commercial radio ("Parish-Pump Radio", 1970):

> …it is depressing to find threadbare arguments being used to justify a development that could have quite *far-reaching effects on the social and cultural life of Britain* (emphasis added).

Although *The Times* still thought that the need for local radio of any sort was doubtful, they argued that if local radio must be developed then the BBC was best placed to provide it, and asserted "the evil effects of the B.B.C.'s radio monopoly have been grossly exaggerated". Commercial services, it averred, would:

> … provide a staple diet of out-dated pop, relieved by snippets of local news – a service that would not be likely to enlarge anyone's range of experience or sense of community involvement.

The new government's White Paper of March 1971, outlining its plans for local "independent" radio, unsurprisingly found little favour in the newspaper's leader column ("A Bad Case for A Bad Policy", 1971). The White Paper was, it declared, "a muddled explanation of a bad policy". It recognised the government's ambition that the stations should be more than "pop and prattle" but argued:

> The essential difficulty over local commercial stations is that there can be little hope of a system that lives up to its apparent cultural pretensions proving commercially viable. If it is to limit its costs and to attract a sufficiently large audience in competition with Radio One and Two, as well as the B.B.C.'s own local stations in a number of cases, then all the evidence suggests that there will have to be large doses of pop music …

always assuming that satisfactory agreements can be reached with the unions on "needle-time", a matter on which there is still a noticeable silence from the commercial interests.

The last paragraph is perhaps the best exposition of the real objections to commercial radio, not least the obviously self-interested argument that it would mean that a significant part of the limited pool of advertising expenditure would be transferred from newspapers to radio:

This White Paper does nothing to remove the fear that *commercial radio in Britain would mean a debasement of broadcasting standards*—especially if the B.B.C. were to go out of their way to compete with the new stations—an *increase in the influence of pop culture...* the whole proposal is based on *diverting advertising revenue from information to entertainment media.* It is ironic that a Government whose very name implies a desire to preserve cultural values should so firmly insist on so undesirable a waste of money (emphasis added).

Some eight months later, as the White Paper was being turned into a bill before parliament, *The Guardian* was dismissive of the urgency with which the government seemed determined to introduce commercial radio – the bill being published just the day after the Queen's Speech at the beginning of the 1971-72 parliamentary session. The paper argued that commercial radio "was not really necessary" ("Not Too Nasty, Not Too Rich", 1971). The main criteria in judging its usefulness was how the services would "enlighten and extend local life". Perversely, it also complained that the bill would result in public service commitments from the new stations under a tight regulator – a "caged tiger even if not entirely tame".

The *Daily Mirror* becomes cooler towards commercial radio

The Labour-supporting *Daily Mirror* clearly had some difficulty following up its editorial line from the 1950s in support of commercial radio, as the debate became highly polarised on a party political basis, with the Labour Party hostile to advertising-funded stations and the Conservatives broadly in favour. Moreover, from 1967, the paper had a new service to champion and defend – the BBC's new service, Radio One. So, the *Mirror* had a slightly schizophrenic editorial line: it was certainly not hostile to the idea of commercial radio but was, on behalf of its readers, nervous that the new stations might mean the end of Radio One.

In addition, proprietor Cecil King was removed in 1968, which may well have led the paper to feel less obliged to promote commercial services.

It had certainly enthused about the launch of the BBC's new "pop" station, with a full-page spread ("Radio One: Here's Your Hey! Hey! Guide", 1967) on the day before the station's launch, which described it a "revolution" and could have come straight out of the BBC's publicity department. The spread included a pithy explanation of the new line-up of BBC national stations (the three existing stations simultaneously changed their names; the Light Programme became Radio Two, the Third Programme, Radio Three, and the Home Service, Radio Four). There was no mention of the "pirate" stations which had led to this "revolution", the controversy over their demise, or to the fact that both the Caroline stations were still on-air. In fairness, the mention of this fact could have led to prosecution under the Marine (etc.) Broadcasting Offences Act.

In March 1969 *The Mirror* reported on what it implied was a clandestine meeting held by what it called "dedicated men" who were drawing up a "battle plan" to make permanent and expand what were still experimental BBC local radio services ("Aunty v. The Big Sell Boys", 1969). The reason for this sense of secrecy and crisis that the paper conveyed was the previous week's pledge by the Conservatives to introduce commercial radio, leading to fears from BBC executives that "they may be nudged out by pressure groups who favour self-supporting commercial radio, broadcasting advertisements". It identified these groups as including the Free Radio Association, the Conservative-controlled Greater London Council (which had gone as far as introducing a bill in the House of Commons to allow commercial radio in London) and, Hughie Green, the TV personality who was to play a big part in the debate about the type of commercial radio to be introduced (but who, in the event, did not apply for the franchises once advertised, as he regarded the terms and obligations too onerous for the stations to be financially attractive). The Conservative policy had led to a counter pressure group named Track (Television and Radio Committee), which the paper said had originally been set up in 1969, being "raised from the grave":

> … with the intention of stopping local radio becoming a 'licence to print money' and to prevent a surfeit of pop music as purveyed by the old pirate stations.

The group had included comedy script-writer Ray Galton, playwright Harold Pinter and Professor Richard Hoggart but, intriguingly, the only current active member was not, as might be expected, a "cultural traditionalist" or liberal-left academic but a London businessman, Mr Barty-King. The *Mirror* said he had toured the BBC's experimental stations "and believes they are setting a gold standard for the development of local radio as a public service". The article implied that there would be a "straight choice" between BBC or commercial local stations – there would not be room for both:

> There will be a battle in July. Local radio listeners will be faced with a straight choice between the B.B.C. and the commercial radio lobby. At present the B.B.C. will have their monopoly and will fight for it... but they may need Mr Barty-King.

Just a week after the Conservatives' unexpected victory (which of course the *Mirror* fought hard to prevent) its front-page lead story (Knight and Desborough, 1970) said that the new Prime Minister, Edward Heath, had:

> ... sparked off a bitter political row ... by ordering full speed ahead for local commercial radio. ... The plan to set up a chain of commercial radio stations will provoke all-out opposition from the Labour Party, who firmly believe that local broadcasting should be a BBC monopoly.

It quoted Heath from a debate in the Commons the previous December, in which the then Opposition leader had said he saw no reason why, if the BBC had local stations, there should not also be commercial ones and further argued that commercial and BBC services could exist alongside in radio, as they did in TV.

Conclusion

The 1945-1973 period was a contested period in British society and politics, in which government control and intervention at every level of economic and cultural activity could either be seen as natural and right, or as unnatural and wrong. With hindsight, the debate in the 30 or so years leading up to the introduction of licensed commercial radio in the UK encapsulates a wider and fundamental battle over the political, economic and cultural direction of the nation. Given that the bulk of research into the country's post-World War Two broadcasting has concentrated on the media effects and political and cultural controversies surrounding television—with radio seemingly regarded as lacking in cultural significance—it is surprising to see how vociferous and heartfelt were the arguments from both proponents and opponents of commercial radio. Radio might have become a secondary or complementary medium after television had established itself as the primary medium for entertainment, information and debate in the evening, leisure and hours, but the fact that radio occupied so many column inches in both the popular and quality press demonstrates that a "background" medium could be, and often was, "foreground" in the public sphere. The newspapers were fully enjoined in this battle and both reflected and promoted the case for and against breaking the BBC's radio monopoly. As can be seen, the "quality press" opposed commercial services ostensibly on the grounds that they would lower the quality of radio broadcasting and produce a malign effect in the broader culture of the nation.

This is not to argue that the newspapers merely had a cynical and disingenuous approach to the topic, but it was certainly convenient for them that Hoggart, Hall and their ilk were able to articulate an intellectual and cultural argument against such services and that these critiques could then be developed and amplified through Leader columns, which in turn would produce more letters and articles, and which then contributed to the debates in parliament and beyond. The *Mirror* had a more nuanced—even contradictory—approach. Its proprietor (until 1968) was alert to the commercial possibilities of investing in both commercial radio and television but he, too, dressed such naked self-interest—and this from a very aristocratic Lord—by attacking the smug and patronising approach of the BBC and the antipathy towards it by much of the paper's working-class readership.

Establishing cause and effect is always difficult in these cases, but it seems likely that the strong support in the "quality press" for the BBC to provide local stations and so extend its radio monopoly encouraged the Labour government of 1966-70 to justify its broadcasting policy in the name of public service, which also provided ideological cover for its antipathy to commercial radio. However, such arguments failed to persuade the Conservative government of 1970-74 which, in its first two years or so, relished its image as a champion of free enterprise. Indeed, it is quite astonishing, given the many serious issues the Heath government confronted in 1970-72 that, right from its first week-end in office, it was keen to stress an early commitment for legislation to establish licensed commercial radio. This enthusiasm may have been prompted in part by a vociferous campaign (both on and off air) for the Conservatives and against Labour by a "revived" *Radio Caroline* during the 1970 general election campaign, highlighting the Conservatives' pledge to break the radio monopoly, and which some leading figures in the Labour party blamed for their unexpected defeat. This argument was given very little coverage in the newspapers; perhaps because they did not want to concede the apparent power of radio to influence public opinion.

The combination, then, of historical accident, the personal characteristics of leading players, sectional interests and fears, and a opposition to popular culture and its associations with "Americanisation" by both the conservative and intellectual liberal-left establishment, played out on the pages of influential newspapers, partially offset by some "cheerleaders" for commercial radio in the tabloid press, helped to produce a unique narrative in the history of broadcasting.

References

A Bad Case For A Bad Policy. (30 March 1971). *The Times*, p.15.

Ariel – Don't Go Back! (1 April 1952). *Daily Mirror*, p.6.

Aunty v. The Big Sell Boys. (12 March 1969). *Daily Mirror*, p.18.

BBC Chief Reviews U.S. Radio. (27 March 1962). *The Times*, p.10.

Choice Of The Manx Wives. (30 April 1965). *The Guardian*, p.19.

Company Meetings – The Daily Mirror Newspaper. (30 May 1953). *The Guardian*, p. 16.

Do the public want it? (3 June 1964). *The Times*, p. 13

Don't Let Caroline Come Ashore. (2 June 1964). *The Guardian*, p.8.

Enough is Enough. (10 May 1968). *Daily Mirror,* p.1.

Hoggart, R. and Hall, S. (18 May 1964). Pirate ships and Realpolitik. [Letter to the editor]. *The Guardian*, p.6.

Hoggart, R. and Hall, S. (31 May1964). Against commercial radio. [Letter to the editor]. *The Observer*, p.30.

Johns, A. (2010). *Death of a Pirate: British Radio and the Making of the Information Age*. New York: W.W. Norton.

Knight, V. and Desborough, J. (25 June 1970). "Free Radio" Hustle By Chataway. *Daily Mirror*, p.1.

Lighter Than the Light. (11 November 1966). *The Times*, p.13.

Local Radio With a Purpose (17 May 1967). *The Guardian*, p.8.

Mealy-Mouthed Arrrogance. (1 April 1952). *The Daily Mirror*, p.1.

Need For National Radio. (15 December1971). *The Guardian*, p.7.

Not Too Nasty, Not Too Rich. (4 November 1971). *The Guardian*, p. 12.

Parish-Pump Radio. (10 September 1970). *The Times*, p.9.

Pop Goes the BBC. (21 December 1966). *The Times*, p.9.

Radio One: Here's Your Hey! Hey! Guide. (29 September 1967). *Daily Mirror*, p.7.

Rudin, R. (2008). Revisiting the Pirates. In S. Nicholas, T. O'Malley & K. Williams (Eds), *Reconstructing the past history in the mass media, 1890-2005* (pp. 115-129). Abingdon and New York: Routledge.

Shearer, A. (26 November 1969). Who Pays for Local Radio? *The Guardian*, p.10.

Sponsored Radio – Readers reply to Viewpoint. (5 April 1952). *The Daily Mirror*, p.2.

The BBC and Its Standards. (3 December 1969). *The Guardian*, p.10

The Toothless Pirate. (23 July 1969). *The Times*, p.9.

CHAPTER FOUR

PIRATE STORIES: RETHINKING THE RADIO REBELS

MATT MOLLGAARD

Introduction

In 1966 a new radio station hit the airwaves in Auckland, New Zealand. Branded as "The Good Guys", Radio Hauraki was a "pirate" commercial radio station, broadcasting contemporary popular international and local music from outside of New Zealand's territorial waters from a dilapidated boat. It was an immediate success for its founders, both in taking audiences from the conservative government broadcasting monopoly and in generating revenue. Radio Hauraki would spend nearly four years at sea, surviving near-sinkings, significant crew and equipment fatigue as well as regular government harassment. After the tough years at sea, Radio Hauraki was granted a licence to broadcast from land. By the time Radio Hauraki came ashore it had paved the way for the significant deregulation of radio broadcasting in New Zealand, a process that would ultimately make New Zealand the most deregulated broadcasting market in the developed world.

The Radio Hauraki story has assumed mythical status in New Zealand broadcasting and amongst the general public, even over 40 years on. At the time, the founders of the station developed considerable public support for the venture by harnessing and mediating young people's antipathy towards the government radio monopoly, which had become complacent about youth audiences due to a lack of any real competition. The founders also created and perpetuated a story of a desperate struggle between a callous and uncaring government and a small band of free-thinking rebels who were trying to create a radio station to play music for the new generation. "The Good Guys" had a readymade and easily recognisable nemesis in the government, making it easy to tap into the rebellious youth spirit of the late 1960s in order to build support for their private commercial radio station.

The discourse that developed during the Radio Hauraki pirate broadcasting saga significantly informed the deregulation of radio broadcasting in 1990 and continues to be influential on New Zealand broadcasting today. This discourse is evident in the most detailed and quoted study of Radio Hauraki's inception, *The Shoestring Pirates* by Adrian Blackburn (1974, 1988). The Radio Hauraki story has been critical in constructing the seemingly polar opposites of public and private broadcasting in New Zealand, which has severely constricted discussion about the role of government and the market in radio broadcasting, resulting in a new stagnation of the radio environment in New Zealand. This stagnation is characterised by a stable and dominant commercial duopoly, a two-channel public system and some minor ethnic, micro and regional broadcasters at the fringes, with no significant new developments for radio broadcasting services in New Zealand for the past decade. While there is still some government involvement in radio broadcasting, the foreign-owned commercial duopoly which now controls around 85 per cent of New Zealand radio (including Radio Hauraki) has managed to maintain a commercial status quo for New Zealand radio services since the early 1990s. This is evidenced in recent government broadcasting policy which does not mention the deregulated radio environment at all (National Party Broadcasting Policy, 2011). The mythologized Radio Hauraki story is part of this dominance as it is now a powerful fable of freedom from government restriction and harassment that speaks to an idealised archetype of New Zealand nationhood; a country that is young, free, classless and progressive. This freedom is not apparent in New Zealand radio with deregulation leading to the consolidation of ownership into foreign companies, formulaic risk-adverse radio aimed at key consumer niches and little real choice outside of the mainstream for audiences.

It is time then to revisit the pirate radio myth, rather than blithely accepting the discourse constructed around it. This deconstruction is necessary to broaden the discussion about broadcasting and how it can develop through public and private initiatives to enhance our lives. Pirate radio has strong associations with recognisable mythological themes— power, rebellion, tribulation and triumph—and so resonates with us on many levels. However, there is another story to be told about pirate radio: the successful businesses, the clever manipulation of public sentiment and the commoditisation of rebellion that are also at the heart of the myth-making around them. These stories are relatively hidden today, but are worth closer examination in order to reflect on what else "The Good Guys" and other pirates achieved.

Pirate radio as a meme

This chapter focuses on the offshore radio pirates of the 1960s. Although there had been illegal broadcasting since what Jesse Walker describes as the "socialization" of radio by governments in the early to mid-1900s, the offshore pirates of the 1960s presented not just a cultural challenge, but an economic and territorial challenge to government controlled broadcasting (Walker, 2001, p. 173). This was critical to the development of the "pirate radio" meme in that the founders of stations like Radio Hauraki were able to tap into this set of cultural understandings held by their target audiences in order to garner support from them and advertisers who wanted to target them. Those radio enterprises that went to sea reinforced their swashbuckling status with their almost outlandish tenacity and their willingness to work outside of society's norms, just like the legendary pirates of the high seas that are so familiar to Western folklore.

Radio Hauraki was part of a wider cultural challenge to state controlled broadcasting in the Western world. The "Reithean" British Broadcasting Corporation (BBC) model of broadcasting heavily influenced New Zealand radio at the time. The Radio Hauraki experience would begin a shift to the market-driven American model. A key difference in the type of pirate operation that arose during the 1960s was whether the state controlled broadcasting on behalf of big business (as in the United States) or on behalf of itself (as in the United Kingdom) (ibid).

It is interesting that the offshore commercial radio pirate was not a significant feature of the American radio system, despite the enormous political and social turmoil of the 1960s. While extra-territorial illegal broadcasting did happen across the border from Mexico and Canada, it tended to be quite different in character from the ship-based pirates of Europe, with its religious or hobbyist programming. While small scale internal illegal broadcasts certainly existed at the time, they were not to challenge the status quo of government broadcasting policy to the extent of the pirates of Europe and New Zealand, who were more focused on breaking their government's advertising monopolies. It seems there were enough entrepreneurial opportunities already available in the United States broadcasting markets to satisfy would-be "rebels" (ibid., pp.197-208).

Michael C. Keith charts the rise and fall of American "underground radio" that was inherently different in character from the offshore pirates of the same time elsewhere. Many "underground" operators took exception to the "pirate" moniker, seeing it as a degrading and politically loaded description of what they believed to be progressive broadcasting (Keith, 2002, pp. 389-404). While American underground radio tied itself to the counter-cultural youth movements of the late 1960s, it was by comparison quasi-legal, becoming a place for experimentation in FM broadcasting after earlier conflicts with the Federal Communications Commission over misusing the AM band. While it could be rebellious in terms of social and political content, its main challenge to societal norms was it development of "free-form" and "progressive" programming styles. The underground radio scene was also not a challenge to a government commercial monopoly. It was essentially entrepreneurial, developing a new commercial broadcasting market on the FM band (Walker, 2001, p. 197).

The European and New Zealand pirate radio experiences of the time were markedly different. Pirate radio in Europe was generally based offshore to avoid very restrictive and punitive broadcasting regulations. Ships from the United Kingdom, The Netherlands, Sweden, Norway, Demark and elsewhere all took to the seas to broadcast popular music and commercials back to their home countries from outside the territorial limits of their government's jurisdiction. On the whole, they were popular and profitable, attracting hundreds of thousands of listeners during a time when much of European (and New Zealand) radio was state-controlled and heavily regulated, even though most of the pirate stations lasted only a few years at best (Walker, 2001, pp. 173-75). Various government strategies such as improving government services for youth (Scandinavian countries) to laws that made marine radio broadcasting a major criminal offence (United Kingdom) helped to kill off the majority of pirates (Yoder, 2002, pp. 135-52). While these stations generally played contemporary music that young people could not find elsewhere on their radio dials, their popularity was also due to the very practice of pirate radio.

Pirate radio practices

Practices are also defining elements of the pirate radio meme. The recognisable themes of the jerry-rigged ships, the danger and the fortitude of the crews as well as the deliberate cajoling of governments and conservative elements of society were essential and exciting elements of offshore pirate radio stations. This coupled with the imported American-style DJ patter on many stations and the popular youth music that was all but banned on government stations was rebellion writ large in an age of increasing youth resistance to the cultural and political norms and accommodations of past (Blackburn, 1988, Yoder, 2002). These practices were political, social and cultural; a challenge to the status quo that resonated with the youth of the time. They also served to disguise the blatantly commercial nature of most pirates as not attempts to make large amounts of money but as progressive broadcasters who promoted freedom in cultural, political, social and economic life. In a sense, pirate radio practice has been fetishised by externalisation of its unusual practices, obscuring the commercial goals most were set up to achieve. The narratives about the pirate ships speak of the desperate nature of radio on the high seas, the ramshackle transmitters, transmission masts and leaky old ships, but also of the heroic efforts to avoid government intervention in order to keep the music playing. The rebellious nature of the DJs, not just as illegal broadcasters but as having "cooler" broadcasting styles and playing new and exciting music for young people also informs this fetishisation (ibid).

These themes are still recognisable today in popular culture, as seen in the movie *The Boat That Rocked* (2009), a comedy about "Radio Rock", a fictitious offshore 1960s radio pirate, apparently loosely based on the experiences of the crew of the infamous British offshore pirate station Radio Caroline. *The Boat That Rocked* is an excellent example of the myth-making power of the radio pirates, the simplified construction of these enterprises as tools of "us" against "them" and the gloss that passing years has added. Alexander Badenoch argues that the film is dishonest in that:

> Setting up this binary conflict makes for a narrative driven by desire for liberation in the form of music and sex. While this conflict serves the film's dramatic aims, it maintains a peculiar silence about the main driving forces of the historical situation: money and law. Except for an oblique reference to the government threatening Radio Rock's advertisers, the film excludes economics entirely from its chains of cause and effect (2009, p. 3).

What *The Boat that Rocked* does not show is that the supposed and idealised mayhem, rebellion and lascivious nature of life on the sea as a radio pirate was an attempt by entrepreneurial and very determined "outsiders" to get what legitimate commercial broadcasters already had – state sanction as well as real and sustainable profits.

However, the film does highlight one interesting facet of the pirate radio experience – the deliberate and conscientious displaying of a version of how pirate radio looks behind the scenes. The ships, the studios, the transmitters as well as the managers and DJs were all photographed, mediated and celebrated in the typical pirate radio narrative. This promotion of the people and the technology was a way of demystifying as well as valorising the pirates—showing their bravery, skill, tenacity and cleverness—and inviting the audience closer to their experiences by mediating them[1]. This externalisation of radio practice is comparatively unusual. Radio practices are typically opaque and mysterious to the listener and at times, even the regulator. This opaqueness is major field of investigation for radio academics. Tim Wall has conducted a study of one British station that uncovers practices of programming that sought to circumvent licence regulations using the ethereal nature of broadcast radio that is actually a construction of carefully managed playlists, scheduling and other practices (Wall, 2006). Alexander Russo (2004) explores the tensions around the authenticity of live radio versus pre-recorded material in early American radio broadcasting, imagined by radio workers as essential to the listeners' trust in the medium. Others such as David Hendy (2005), Martin Shingler & Cindy Wieringa (1998) and Andrew Criswell (1994) and in this volume Helen Wolfenden and J. Mark Percival have also explored radio's usually unseen conventions, rhythms and practices. Pirate radio reversed this opaqueness of practice by promoting the devices and work patterns that got them to air, making them seem more accessible to their audiences and reinforcing the dangerous and difficult nature of the broadcasts they were risking all to provide for their eager audiences.

Radio Hauraki as a pirate radio station

Radio Hauraki, taking its cue from Radio Caroline, exemplifies the pirate radio meme. From its inception as an attempt to break the New Zealand government's monopoly over commercial broadcasting revenue,

---

[1] Good examples of collections of photographic and other visual material from the time include Paul Harris (1968-2007), *When Pirates Ruled the Waves* and John Monk (2007), *Radio Hauraki: The Pirate Years 1966-1970.*

its owners grew a support base by publicity of attempts to obtain a ship, fit it out and sail it out of New Zealand waters to broadcast popular youth music not heard on government stations offshore of New Zealand's main city, Auckland. Those behind the radio station included former government broadcasters who saw the commercial potential of capturing some the enormous income of the government's monopoly and set about trying first to gain a legal land licence. After that was denied, the decision was made to take the fledgling station—that was already generating considerable interest from potential advertisers—to sea (Blackburn, 1998, Monk, 2007). The founders skilfully used the media in promoting their cause, tipping them off to meetings with government ministers and public addresses that provided dramatic copy for the news media. They took this media handling to new levels when "leaked" news of their ship making ready to sail encouraged public participation in illegal attempts to leave shore and obstruct police and other officials (Blackburn, 1988).

The two ships that Radio Hauraki broadcast from, *Tiri I* and *Tiri II* went on to become became household names in New Zealand. The various dramatic equipment failures, storms, groundings and other tribulations that Radio Hauraki suffered offshore were front page news in New Zealand, with attendant pictures of the stricken ships and the weary "pirates" and their supporters (Blackburn, 1988, Monk, 2007). The pirates also made a virtue out of the pre-recording of almost all on-air material days and weeks before it was to be broadcast. Listeners were invited to the land-based recording studios to watch programmes being recorded and pictures were circulated of the dramatic "drops" of programmes from a light plane to the boats (Monk, 2007, p. 118). This transparency of practice was in stark contrast to the impenetrable nature of the government broadcasters, whose studios and offices were not open to the public to anywhere near the same degree and whose broadcasters were more removed and "professional", trained in "correct" Received English pronunciation and "acceptable" broadcasting styles, largely inherited from the BBC. Compared to the raucous, rebellious and dashing pirates of Radio Hauraki, the government broadcasters were generally aloof, cautious and unavailable to the listeners, further contrasting the public and private services.

The pirates were heavily promoted as "The Good Guys", with promotional pictures taken on the decks and in the studios of the ships and American-style DJ profiles in newspapers, that gave details of hobbies, thoughts about being a pirate as well as preferred cigarette brands and lists of favourite popular artists (Monk, 2007, pp. 51-3). The founders of the station become media stars in their own right, appearing in national newspapers and in front of crowds of jubilant supporters and journalists to make their case (Blackburn, 1988). This strategy juxtaposed the open, liberal, rebellious and progressive private radio raconteurs with the staid, conservative and culturally stifling government owned radio stations. This engendered the pirates to the nascent youth market, who were beginning to pick up counter-culture cues from the tumultuous experiences of youth cultures in other Western countries, most notably the United States and the United Kingdom (Blackburn, 1988, pp. 70-6). The Radio Hauraki pirates and their experiences also resonated with wider memes in New Zealand society, such as that of the "Man Alone", the self reliant battler against the system of author John Mulgan (1939) and the "can do" attitude that supposedly informs the tenacity and ingenuity of New Zealanders and their enterprises.[2]

However, behind the open displays of practice, the rebellion, the boats and the pirates, the Radio Hauraki story is a significantly commercial one. The founders were determined to break into the government broadcasting advertising monopoly, one way or another. The tenacity displayed in getting to air and staying there was informed chiefly by the desire to make a large amount of money. Even before the boats left port, large corporate advertisers had been secured to be on-air from day one (Blackburn, 1988, pp. 26-30). The extra-jurisdictional nature of offshore broadcasting also opened up advertising avenues for products that were banned from onshore media during certain times of the day, such as tobacco and later alcohol (Monk, 2007, p. 80). Notably, when the original company that owned Radio Hauraki was wound-up after coming ashore, it managed to post a loss of NZ$5, therefore never paying tax, despite making thousands of dollars from advertising (Blackburn, 1988, p. 158). This element of Radio Hauraki's commercial radio practice was and has remained relatively unexplored. The struggle to get Radio Hauraki to air overshadowed debates about the political and economic challenges to New Zealand broadcasting that Radio Hauraki also represented.

The David versus Goliath struggle that was portrayed by the pirates and enthusiastically supported by Radio Hauraki fans and some media had the effect of chilling the debate about broadcasting services in New Zealand to this day. There are practically no significant critiques of the Radio Hauraki challenge to broadcasting development in New Zealand.

---

[2] A good general introduction to New Zealand and its people can be found in Michael King (2003), *The Penguin history of New Zealand*.

Examples of this can be seen in supposedly neutral sources such as *Te Ara: the Encyclopaedia of New Zealand*, a government publication which presents the story with very little context and no critique at all:

> In 1966 Radio Hauraki declared it would challenge the state's commercial radio monopoly by transmitting from a "pirate" boat off Great Barrier Island. As the vessel left port, police boarded the boat, disabling its engine and arresting the crew. The pirates soon tried again, and succeeded. In 1970 the government bowed to public pressure and legalised private radio stations. (Popular Culture, n.d.)

This summary is also exemplified and amplified by the contemporary owners of Radio Hauraki:

> There once was a time when the New Zealand Government decided what would be on the radio. In 1966 a few good men set out to change that. The MV Tiri had a transmitter installed and was anchored three miles offshore in the Hauraki Gulf, just outside Government jurisdiction. Despite the best attempts by Government (and mother nature), New Zealand's first private radio station was born. In 1970, the Government finally caved and allowed Hauraki to broadcast legally from land, breaking their long-held monopoly over the airwaves. This started the entire private radio industry in New Zealand. Hauraki has been playing great music ever since. (Our History, n.d.)

The years of Radio Hauraki's rebellion served to polarise arguments about the role of the media in New Zealand. Essentially, government services became "bad", while private operators became "good". When Radio Hauraki was granted a land licence after 1111 days at sea, earlier interest in the station had waned, but its legacy was a strong and virtually unchallenged belief in the rights of private enterprise in broadcasting (Blackburn, 1988, pp. 158-60). The government of the time came to accept the "inevitability" of private radio and tried to avoid further rebellion against its policies by facilitating the private sector in getting licences to broadcast through a Broadcasting Tribunal (ibid., pp. 149-154). In one sense, Radio Hauraki had "won", although the victory can be seen as also about the realities of the election cycle during an era of generational and economic challenge and change. There is a sense of the government growing both weary and wary of the Radio Hauraki situation, with an eye to parliamentary elections and a new generation of younger and worldlier voters pushing politicians to liberalise aspects of New Zealand's previously conservative consensus (ibid.). The Radio Hauraki experience can be seen as another tranche of this steady opening up of the political system to new ideas during the late 1960 and early 1970s.

Despite this, broadcasting policy in New Zealand has never since enjoyed the visibility of the Radio Hauraki years. The slow but steady liberalisation of the radio regulation continued until the virtually total deregulation of the 1990s, brought about by a significant reorientation of the New Zealand economy towards the free market and global capital (Kelsey, 1997). Radio Hauraki had laid the groundwork for media deregulation by opening up of the radio spectrum to private enterprise. Further deregulation was seen only as a matter of degrees and not particularly contentious (Duignan & Shanahan, 2005). Since the struggles of the Hauraki years, broadcasting has become a somewhat pariah government portfolio and is typically a secondary burden for a busy cabinet minister with more "important" portfolios to manage.[3] Debates about commercial radio broadcasting in New Zealand would seem to be "settled".

The effect of Radio Hauraki in context

The conversion of a media system from a largely public asset to a largely unfettered private market has profound implications. The deregulation of New Zealand radio has produced political, economic, cultural and social outcomes that are unique to New Zealand broadcasting, but also reveal more universal facets of capitalist media structures (Golding and Murdock, 1997, p. xvi). As the mass media are "fundamentally economic institutions" the critical issues of ownership, management and institutional structures of the mass media affect the products they produce (Gomery, 1989, p. 43). Key to this analysis is the most basic element of the organisation of the mass media – who owns it. Douglas Gomery (1989) argues that development of the mass media has created an oligopoly that needs to be interrogated as an economic phenomenon in terms of its structure, conduct and performance. In New Zealand, debate about radio broadcasting systems, their value and their economic and cultural worth have struggled to mature beyond the polarisation of the market versus the government model, with Radio Hauraki used as the prime example of the power and the utility of market-driven models in providing choice. This is reflected in the highly commercialised New

---

[3] For example, as of late 2011 the Minister for Broadcasting is also the Minister for Commerce, Associate Minister for Education and Associate Minister for the Accident Compensation Corporation.

Zealand broadcasting market of today and the lack of any new government radio broadcasting operations for many decades.

This is in contrast to the outcomes of pirate radio in Europe. Despite the real challenges to the status quo provided by pirate radio in the United Kingdom, The Netherlands and the Scandinavian countries, it generally served to bolster the debate about the appropriate mix of radio services, leading to more government services for more people and private radio with regulatory requirements to fulfil needs for different audiences (see Rudin, above). The United Kingdom responded to the pirate challenge by revisiting government broadcasting services to broaden them while continually developing and revisiting regulation. The United Kingdom used the challenges of the offshore pirate years to inform supervision of both public and private radio in order to meet wider goals of broadcasting plurality and diversity (Hitchens, 2006, pp. 12-18). Broadcasting is a major area of public contention and policy development in the United Kingdom and its attendant portfolio is an important government post (ibid., pp. 86-102).

The real and lasting legacy of Radio Hauraki can be seen in the current New Zealand radio market. The successful fight for a commercial radio licence for Radio Hauraki is widely recognised as the starting point for the deregulation of radio in New Zealand, even if there was a hiatus in granting new commercial radio licences between 1972 and 1989 when full deregulation began (Johnstone, 2005, p. 49, Dubber, 2007, pp 35-6). Today, the New Zealand radio market is the most deregulated in the developed world with significantly more radio signals per capita than anywhere else. There is a signal for approximately every 5200 people (Duingan & Shanahan, 2005). The effect of the deregulation of radio broadcasting in New Zealand is a market that is dominated by two foreign owned companies who claim around 85 per cent of all frequencies and also radio audiences in New Zealand. With the small public system and a few struggling broadcasters at the fringes, New Zealand radio can be characterised as highly commercial, formulaic, risk adverse and dominated by two extremely competitive commercial companies with small players unable to significantly challenge the commercial duopoly for audience share or commercial gain (Mollgaard, 2010, Duingan & Shanahan, 2005). The concentration of ownership of New Zealand radio demonstrates "the new oligopoly" model of Roger Wallis and Krister Malm that is the result of deregulation without policies that provide for diversity of ownership and content, access and localness (Malm & Wallis, 1993). New Zealand radio has become a comfortable commercial consensus, with little opportunity to challenge the entrenched norms of free market economics dominating New Zealand's radio services.

Conclusion

Somewhat ironically, the foreign multinational company that bought the New Zealand government's commercial radio services when they were sold in the 1990s also owns the contemporary Radio Hauraki. The radio station would be unrecognisable to its earlier supporters. It now broadcasts a strictly regimented nationwide networked rock music programme, from deep inside the bowels of a highly secure building in the Auckland CBD. The rebellion and transparency of practice of the early Radio Hauraki has all but gone. Instead, a highly regimented and formatted programme, featuring highly paid and market researched personalities and a small and targeted playlist of only the most popular mass-appeal rock songs is generated in a secure studio far from prying eyes (Mollgaard, 2007, The Radio Network, 2011). There is no room for rebellion beyond a manufactured "rock attitude" that is designed to capture the 25-50 blue-collar male listeners. Radio Hauraki is a product for the mass market, produced by a "radio factory" that earns revenue by networking eight highly targeted mass-market formats throughout New Zealand. This is the ultimate outcome of a fully commercial market – radio stations that must attract specific niche audiences (and keep them) in order to make enough money to return profits to shareholders who can invest their money anywhere else if they so wish. Ultimately, nothing else but profits can *really* matter in this system. And perhaps, for Radio Hauraki and the bulk of the 1960s offshore pirates – nothing else ever *really* did.

Radio Hauraki not only provided new music, styles and energy to radio broadcasting in New Zealand, but also forced the government to rethink their approach to youth audiences. It was also very exciting. There was danger, drama and bravery – mixed in with rebellion and the changing of the cultural milieu during the 1960s. The government's solution was to first try to prevent the commercial challenge to its radio monopoly, then ignore it, then facilitate it, albeit, on its own terms. However, this was not enough to insulate New Zealand broadcasting policy from the encroachment of the market. The fact that Radio Hauraki was able to "come ashore" and begin legitimate broadcasting is critical in that the story of Radio Hauraki became a parable about a successful rebellion and practically unchallengeable – a powerful myth that is still celebrated today.

But, ultimately, it was a rebellion that was co-opted by commercial interests. The real legacy of Radio Hauraki is the ascendance of formulaic private commercial broadcasting – mostly for the benefit of foreign shareholders and multinational companies, but also a lack of discursive space in which to challenge the highly commercial broadcasting economy of New Zealand. Radio Hauraki's subversion of the 1960s seemed to promise a new era of open and progressive radio services for New Zealanders. In reality Radio Hauraki broke the government radio monopoly, but it also helped to create a new stifling, corporate and highly commercial status quo.

References

Badenoch, A. (2009). *The peculiar silence of "The Boat That Rocked"* Zeitgeschichte-online. Retrieved 16 July, 2011, from http://www.zeit geschichteonline.de/portals/_rainbow/documents/pdf/badenoch_boat.pdf.

Blackburn, A. (1974 & 1988). *The Shoestring Pirates*. Auckland: Hauraki Enterprises.

Criswell, A. (1994). *Understanding radio*. London: Routledge.

Dubber, A. (2007). McLuhanising radio: essaying a media ecology approach to technological shift in New Zealand radio broadcasting' in *The Communication Journal of New Zealand/He Kohinga Korero*. 8 (1), 23-37.

Duignan, G. and Shanahan, M. W. (2005). The impact of deregulation on the evolution of New Zealand commercial radio. In Neill, K. & Shanahan, M. W. (Eds.), *The great New Zealand radio experiment*, Thomson Dunmore Press: Southbank, pp. 17-46.

Hendy, D. (2000). *Radio in the global age*. Polity Press: Cambridge.

Golding, P. & Murdock, G. (1997). Introduction: communication and capitalism' in Golding, P. & Murdock. G, (Eds.), *The Political Economy of the media I*. Cheltenham: Edward Elgar Publishing, pp. xiv-xviii.

Gomery, D. (1989). Media economics: terms of analysis. *Critical Studies in Mass Communication*, 6, (1). 43-60.

Harris, P. (2007). *When pirates ruled the waves*. (6th ed.). Glasgow: Kennedy and Boyd.

Hitchens, L. (2006). *Broadcasting pluralism and diversity: a comparative study of policy and regulation*. Hart Publishing: Oxford.

Kelsey, J. (1997), *The New Zealand experiment: A world model for structural adjustment*. Auckland: Auckland University Press.

Johnstone, B. (2005). Changing the game - entrepreneurs bring networks to small markets. In Neill, K. & Shanahan, M. W. (Eds.), *The great*

New Zealand radio experiment. Thomson Dunmore Press: Southbank, pp. 47-66.

Keith, M. C. (2002). Turn on ... Tune in: the rise and demise of commercial underground radio. In Calhoun, C. (ed.), *The radio reader: essays in the cultural history of radio*. Routledge: New York, pp. 389–404.

King, M. (2003). *The Penguin history of New Zealand*. Auckland: Penguin.

Monk, J. (2007). *Radio Hauraki: the pirate years 1966-1970*. Publishing Press Ltd: Albany.

Malm, K. & Wallis, R. (1993). From state monopoly to commercial oligopoly: European broadcasting policies and popular music output over the airwaves. In Bennet T., Firth S., Grossberg L., Shepard, J. and Turner, G. (Eds.), *Rock and popular music: politics, policies, institutions*. London: Routledge, pp. 156-68.

Mollgaard, M. (2005). Can a multi-national care about the Kiwis? CanWest and Kiwi FM, in *Radio in the world: papers from the 2005 Melbourne Radio Conference*. Healy, P., Berryman, B. and Goodman, B. (Eds.). Melbourne: RMIT Publishing. pp: 225-235.

Mollgaard, M. & Rosenberg, B. (2010). Who owns radio in New Zealand. *Communication Journal of New Zealand/He Kohinga Korero*. 11 (1). 85-107.

National Party Broadcasting Policy 2011. Retrieved October 22, 2011, from http://www.national.org.nz/PDF_General/Broadcasting_policy.pdf

Popular culture: Auckland Region (n.d.). *Te Ara –the Encyclopaedia of New Zealand*. Retreived October 12, 2011, from http://www.teara.govt.nz/en/auckland-region/13

Radio Hauraki: Our History (n.d.). Retrieved October 12, 2011, from http://www.hauraki.co.nz/our-history

Russo, A. (2004). 'Defensive transcriptions: radio networks, sound-on-disc recording, and the meaning of live broadcasting. *The Velvet Light Trap*. 54 (3), pp. 4-17.

Shingler, M. & Wieringa, C. (1998). *On air: methods and meanings of radio*. Oxford University Press: New York.

The Radio Bureau, (n.d.). *History of the New Zealand radio market*. Accessed 16 July 2011.

Walker, J. (2001). *Rebels on the air: an alternative history of radio in America*. New York University Press: New York.

Wall, T. (2006). 'Calling the tune: resolving the tension between profit and regulation in commercial radio'. *Southern Review*. 39 (2), pp. 77-95.

Yoder, A. (2002). *Pirate radio stations: tuning into underground broadcasts in the air and online*, McGraw-Hill: New York.

CHAPTER FIVE

WIDE OPEN ROAD:
RADIO AS CULTURAL HISTORY

TOM MORTON

In November 2008 the Australian Broadcasting Corporation's national youth radio network JJJ broadcast a series of four one-hour radio features entitled *Wide Open Road*. The series - named after the 1986 single of the same name by Perth band The Triffids - set out to explore the relationship between Australian popular music and the landscape. In particular, it aimed to tell a story. As the narrator, Richard Kingsmill, says at the beginning of the first episode, it's the story of "how a generation of Australian musicians discovered the landscape they live in through music."

This series of radio features was broadcast in prime weekend listening time, at 2pm on Saturday afternoons over four weeks. For JJJ, it was, in some senses, a bold programming move.

Although JJJ had played music features and documentaries in its early years as a national network, and had certainly done so in its previous incarnation as Sydney station JJ, at the time *Wide Open Road* was broadcast the vast majority of the network's output was presenter-driven music radio with a strong focus on the 16-25 demographic. Highly-produced radio music features were thus a new and unusual listening experience for the majority of JJJ's audience. To guide them through this experience, the producers of the series chose one of the station's veteran music presenters, Richard Kingsmill. Kingsmill is the station's music director, currently hosts its Sunday afternoon new releases show, and is widely recognized as an authoritative voice on Australian rock music.

The series was also unusual in that it was broadcast simultaneously on ABC Radio National, the ABC's cultural radio network, whose core listenership is generally much older than JJJs; typically in the 45 – 65 age bracket, or older.

Wide Open Road was a largely unprecedented co-production between both networks. As such, it consciously sought to appeal to, and create a bridge between, two quite different audiences or "musical taste-publics", as Paddy Scannell has called them (Scannell, 2009, p. 89). Broadly speaking, the experiment appeared to be successful.[1]

The series of radio features was accompanied by a website, www.abc.net.au/wideopenroad, which featured extended interviews with many of the musicians who appear in the radio series, a "lyrical map" of Australia and other interactive elements. It also allowed visitors to the site to stream the radio features (they could not be podcast due to issues with music rights).

Wide Open Road was researched, written and produced by myself and another veteran documentary producer, Cath Dwyer, who had also worked at JJJ. At the time, I was a full-time radio feature and documentary producer with the ABC. In 2010 I became a full-time academic with research interests in radio studies and investigative journalism.

In the following discussion, I approach *Wide Open Road* as a radio essay in cultural history; a non-linear, thematic exploration of the history of Australian popular music. In the first part of my discussion, I argue that it exemplifies some of the distinctive ways in which radio can embody a cultural history of sound, voice and performance. In the latter part, I reflect on what was missing from both the radio features and the website; a recognition of the important role that *radio itself*–and in particular, community radio–played in the shaping of this cultural history.

The thesis of the series is simple and bold. Before the mid-1970s, Australian songwriters working in rock or pop music wrote songs as though they lived somewhere else – usually in the United States. The first hit single for Australian band Axiom, released in 1969 and written by Brian Cadd, was titled *Arkansas Grass*. Cadd's subsequent breakthrough

---

[1] One comment posted by a listener on 21 November 2008 is indicative: "Great series, can't say I listen to jjj much but I enjoyed this, a real education for me as I hadn't heard of a lot of those old bands before and a trip down memory lane for my ex rock chick Mum".*Wide Open Road* Forum. Retrieved April 14, 2011 from http://www2b.abc.net.au/tmb/Client/Board.aspx?b=145

single as a solo artist in 1972, *Ginger Man*, has the narrator—possibly a Vietnam veteran—travelling through time and space across America.

From the mid-1970s onwards, something happened in Australian popular music. Beginning with the successful pop band Skyhooks, whose songs are replete with references to places in suburban Melbourne, Australian songwriters moved away from simply imitating the iconography of American popular music, and began to explore the physical and psychic geography of their own country.[2]

In other words, they began to write songs as though they lived in Australia: songs set in an Australian landscape, with Australian place names, and which evoke the sights, smells and sounds of the city and country.

The radio series sets out to explore how and why that particular cultural shift happened, and what possibilities it opened up for the creation and exploration of new Australian cultural identities, through interviews with more than fifty songwriters and musicians, archival concert recordings, and—of course—the music itself.

The four episodes of the radio series are titled *Road*, *Suburbs*, *Carpet*, and *Coast*, and the website, too, is structured around these four words. The titles were chosen either because they refer to recurring images or metaphors in the songs, or to the places in which the songwriters situate themselves. *Road* is about imaginary—and sometimes literal—journeys through the Australian landscape. *Suburbs* is about the urban environment, about the places where most of the songwriters come from, and the naming of place in their songs. *Coast* is about the interplay of coastal culture and popular music, from the surf bands of the 60s to the transnational "roots" music of the late 1990s and early 2000s. In a sense, it is also about the Australian sublime in popular culture. *Carpet* is, as Tony Mitchell has written in a review essay on *Wide Open Road*, "a synecdoche for music venues, mostly pubs" (Mitchell, 2009). In other words, the titles

---

[2] There is one very significant exception to this broad trend, namely country music, and Aboriginal country music in particular. As Clinton Walker shows in his groundbreaking book *Buried Country* (Walker, 2000), and the accompanying documentary film, Aboriginal musicians like Dougie Young, Herb Laughton, and Bob Randall were writing and performing songs unequivocally set in Australia from the 1940s onwards. This is no less true for non-indigenous country songwriters like Tex Morton or Slim Dusty.

refers to the sticky, beer-and sweat-soaked carpet audiences stand on when they are watching a band; so *Carpet* is a history of Australian pop music as live performance.

Perhaps unusually for a conventional radio series, the four episodes are not designed to be heard in sequence. The series is not a single chronological history, but four parallel narratives which each illuminate an important moment in Australian cultural history with different accents and from different perspectives.

What is this historical moment – and why is it important?

Put very simply, it's the moment in which Australian songwriters begin to articulate a distinctively Australian voice; and to develop a distinctively Australian sound in popular music. As Clinton Walker puts it in his book *Stranded: the Secret History of Australian Independent Music 1977-1991*, it is the moment in which Australian popular culture comes of age (Walker, 1996, p. viii).

This is a large claim to make. In my view, however, Walker is right in identifying a real and significant shift, not just in the expressive possibilities of Australian pop music, but in Australian cultural identity itself. It is, moreover, a shift which most other cultural historians have missed, ignored, or under-emphasized. In his broad and otherwise authoritative survey of Australian popular culture *Making it National: Nationalism and Australian popular culture*, Graeme Turner argues that the 1980s were "a decade of revived Australian nationalism" (Turner, 1994, p. 3). As Turner describes it, the Australian national character celebrated in the 1980s was "prescriptive, unitary, masculinist and excluding" (ibid).

Yet Turner devotes only seven out of one hundred-and sixty pages to pop music, and those entirely to one band, Yothu Yindi. Turner is by no means alone in down-playing the role of music in Australian popular culture. In the 1980s and 1990s, Australian cultural studies academics generally paid scant attention to pop music. The foundational texts on the history of Australian popular music were written by music journalists working outside the academy. In the last decade, however, a lively and wide-ranging body of academic writing about Australian popular music and its many sub-genres has emerged; but, until very recently, little of it was concerned either with history, or the relationship between pop music and locality.[3]

The reader could, perhaps, be forgiven for wondering at this point whether what she or he is reading might better have been published under the rubric of popular music studies, or cultural studies, rather than radio studies. What I want to suggest, however, is that *Wide Open Road* is an illustration of how radio – and in this specific case, the radio feature or documentary – can function both as a mode of cultural enquiry, and a distinctive way of narrating cultural history.

Wide Open Road can be read—or as I suggest, heard—from a range of different perspectives: as an exploration of the relationship between popular music and locality; as an essay in oral history; as a document of one aspect of the developing discourse of Australian national identity in the 1970s and 1980s. It exists on the borderlands between a number of existing fields of academic inquiry. In other words, it could be said to be a piece of multidisciplinary cultural history; and, as I shall argue, a piece of *embodied* cultural history. As such, it seems to me to represent an example of the particular contribution which Kate Lacey has argued that radio, and the study of radio, can make to historical writing about media and culture: that is, to "amplify the 'blind spots' of visual cultural histories and help to sound out the connections and inter-relationships between the full range of communicative practices available in any particular time and place" (Lacey, 2008, p. 30).

On the basis of our interviews with around sixty songwriters and musicians we attempted to explore the nature of their relationship with the landscape, their sense of place and belonging, and their understanding of themselves as artists operating within a distinctively Australian cultural context. Out of this exploration emerges a very different version of the cultural history of the 1980s to the one described by Turner - "prescriptive, unitary, masculinist and excluding." From the late 1970s onwards, popular music began to articulate a range of counter-narratives to the dominant discourse described by Turner; a hybrid version of Australian national identity which was—to use Turner's own formulation—"a creative,

---

[3] For an excellent introduction to recent writing about Australian popular music, see *Sounds of Then, Sounds of Now. Popular Music in Australia* (Homan & Mitchell, 2008). For a more general survey of the relationship between popular music and place, see *Sound Tracks. Popular Music, Identity and Place.* (Connell & Gibson, 2003), and *Music, National Identity and the Politics of Location. Between the Global and the Local.* (Biddle & Knights, 2007)

resistant, cultural and political process of becoming – rather than a conservative, already completed project of exclusion" (Turner, 1994, p. vi).

This level of analysis isn't made explicit in *Wide Open Road*, which was produced, quite consciously, for a popular audience, though it is there in the subtext. What is made explicit is that, in popular music, the historical moment out of which these counter-narratives emerge is also the historical moment of punk rock. When Brisbane band the Saints released their debut single *(I'm) Stranded* in late 1976, they kicked open a door which a whole generation of Australian musicians would walk through. As Nick Cave says in *Road*, (in an archival interview recorded in 1991):

> The Saints had a genuine gut level contempt for everything going – it was very Australian, it was very different from the English punk thing. (I'm) Stranded gave me and a lot of people around me a kind of soundtrack to the way we wanted to live, it gave us a licence to behave in a certain way.

In other words, punk opened up a new cultural space.

Punk gave the generation of musicians who came of age in the mid-late 1970s a freedom to experiment, sonically and lyrically; a freedom which they would use to begin to explore their own culture, their sense of identity and sense of place. One of the paradoxes we explore in the series is that this exploration was often expressed in deeply negative terms: as a profound sense of isolation and alienation from Australian culture and from the landscape itself.

So, for example, David McComb, songwriter and lyricist for The Triffids, on the 1986 album *Born Sandy Devotional*, delivered a suite of songs which are psychodramas of lost love and betrayal set against the vast spaces of the West Australian desert and its meandering coastline. There's a kind of brooding quality to the musical landscape of these songs – a sense of grandeur – but also menace. Or as Steve Kilbey, songwriter for The Church and a passionate admirer of McComb's puts it in *Road*, McComb was

> ...one of the first guys to start getting an Australian Gothic into songs, the horror of it, the Gothic quality, not a dark snowing town in Germany, but this wide open space and the blazing hot sun... I think McComb really got that colonial white man's horror.

By contrast, in *Cattle and Cane*, The Go-Betweens' Grant McLennan evokes the farming countryside of Queensland with a poignant lyricism, tinged with nostalgia for the remembered landscapes of childhood. McLennan died in 2006, before we commenced interviewing for *Wide Open Road*. A contemporary of McLennan's, Paul Kelly, who is himself one of Australian pop music's most prolific and widely-admired songwriters, described *Cattle and Cane* as follows:

> It was one of those songs that you could see and smell, you could smell Northern Queensland in it, the cane, his father's watch in the shower...it just had...just telling details that somehow open the song up, make it richer.

In the episode *Road*, listeners are invited to consider a paradox; that the songwriters who initiate this imaginative discovery of the Australian landscape are predominantly "white suburban boys", to use Clinton Walker's phrase (Walker, 1996, p. vi), many of whom have spent little time outside of the cities. In the latter part of the episode, we tell the story of how some of these white suburban boys—members of the band Midnight Oil—toured remote Aboriginal communities in the Northern Territory with the Warumpi Band on the "Blackfellah Whitefellah Tour", and the profound effect that this experience had on their songwriting, and the sound of their music – described here by drummer and principal lyricist for the band, Rob Hirst:

> We came back with our acoustic guitars and our drums and our drum sticks and our hair all smelling of campfire, and I think before it was even washed out of our hair and out of our clothes we started writing these songs inspired by the desert, and we started writing really simple songs that would have sounded good out there. We called them campfire songs, because when we sat down around a campfire and we asked the folks out there to play their music with the clap sticks and at the Top End with the didge but mainly with just vocals, just with voices, under those huge stars and in what Charlie McMahon used to refer to as 'the great quiet', it sounded so potent and so appropriate, and our big city white fella music didn't.

In the mid-1980s, Midnight Oil's music, and the band's public rhetoric, was deeply engaged with Aboriginal struggles for self-determination and the hand-back of Uluru (formerly known as Ayers Rock) to its traditional indigenous owners. The actual and imaginative journey which Rob Hirst describes above, and the songs it produced, represents one counter-narrative to the dominant discourse of Australian national identity described by Turner. Others are explored in the remaining episodes of the radio series and in the individual episodes featured on the website.[4]

If, as I've suggested, *Wide Open Road* offers a new and original perspective on an important moment in Australian cultural history, the question still remains as to what, if anything, is distinctive or unique about the medium in which it is offered. To put it more succinctly: how is *radio* as cultural history different? In what ways does radio, as a medium and mode of enquiry, allow us to experience a field of cultural history—in this case, Australian pop music—that a book, a journal article, or a TV series might not?

One possible answer to these questions is that, at least in relation to the printed word, radio can function not only as a way of narrating the cultural history of popular music – but as a popular history of culture.

Wide Open Road was conceived, written and produced for a popular audience, at least so far as the audience for the youth network JJJ was concerned. A large section of JJJ's audience was neither born nor thought of when much of the music in the series was produced; as such it was an attempt to introduce a new generation of young Australians to their own cultural history – their own tradition, one might even say.

This particular answer, however, does not take us very far. I want to suggest that a more interesting way to think about radio as cultural history is to think of it as an *embodied* history – a history of voice, sound and performance.

The critical and theoretical literature on embodiment is rich and complex, and spans a number of academic disciplines. My thinking has been informed in part by Paddy Scannell's exploration of the grain of the radio voice in his writing about Vera Lynn. Scannell explicitly refers to Roland Barthes' famous formulation - "the grain of the voice, the body in the voice as it sings", arguing that "the effect of the studio microphone... was to repersonalize the voice" (Scannell, 2008, p. 394). For Scannell, Vera Lynn was one of the first radio performers (in Britain at least) to understand and learn to inhabit the intimacy and sincerity which the studio

---

[4] So, for example, *Suburbs* touches on the expression of non-Anglo-Australian identities in punk rock by The Hard-Ons and in Australian hip-hop by TZU and others, while *Coast* explores the emergence of a hybrid coastal music.

mike afforded. Her radio voice displaced the "public, performative values" of the traditional singing voice, and substituted for them "private performative values that privileged the particular and the personal" (Scannell, 2008, p. 395).

In a very different context, Virginia Madsen explores the "incarnate" radio voice in the work of the Australian feature and documentary maker Kaye Mortley. In Mortley's radio works, argues Madsen,

> ...voices never appear disincarnate, but carry traces of the time, flesh and breath that carried them...the voice offers a *revealing* of the body as it is carried by the voice and released into the exterior world" (Madsen, 2009).

The voices of Australian rock singers and songwriters featured in *Wide Open Road* are – quite literally – a world away from the voice of Vera Lynn or the characters in Kaye Mortley's radio features. But these voices too, carry "traces of the time, flesh and breath that carried them". One of the most distinctive and idiosyncratic of those voices belongs to Dave Graney. Graney was born and raised in Mt Gambier, a town of some 12,000 people in the south-east of South Australia. After leaving school in the mid-1970s, Graney moved to Adelaide and formed The Moodists with Claire Moore, the band's drummer and Graney's long-time partner and musical collaborator. Like many of their musical contemporaries, the Moodists left Australia in London in the early 1980s. In *Road*, Graney reflects on this period of self-imposed exile, and his relationship with Australia at the time:

> I was not Australian. I had no idea of ever returning to Australia. There were things you could draw on in Australian culture, but there wasn't and isn't much. In our band we loved to talk about movies like *Wake in Fright*. It is a very gothic film. Donald Pleasence is a malevolent character in it, drinking stale, fly-blown beer, eating rancid kangaroo testicles, and Chips Rafferty is this horrible presence, he just keeps going, "Do ya want a beer?" "No." "What are ya? What's the matter with ya?"

Here Graney caricatures, and quite consciously rejects, the 'masculinist' version of Australian identity celebrated in Australian popular culture in the 1980s according to Turner. In the 1990s, Graney himself did return to live in Australia, and on the album *Night of the Wolverine* produced a number of songs replete with references to his childhood in Mount Gambier and the back-roads of the countryside around it.

What is revealing about the excerpts from the interview with Graney in the radio series, however, is that, in the very moment that Graney is disavowing any sense of Australian identity, his voice affirms it. There's something in the grain of both his speaking and his singing voice – a deep baritone, laconic in its delivery, ironic in its attitude to the many personae which Graney himself takes on in his songs - which bears witness to the place and the historical moment from which it emerges. It is an unmistakeably Australian voice, steeped in the very experiences of country-town life from which Graney has distanced himself, both physically and imaginatively.

One further example, which also happens to involve a South Australian musician, may help to illuminate further the claim I'm making here: that radio offers us (by contrast with text or television) an embodied history which asks us to pay attention to the voice.

The first part of *Road* explores the imagery of the road and the road trip, both real and imaginary, in the songwriting of a number of non-indigenous musicians, most of whom began writing in the late 1970s and early 1980s. The second part of the feature gives this narrative a twist, when it begins to explore an alternative musical history from the "wrong side of the road". The phrase is taken from the title of a feature film, *Wrong Side of the Road*, released in 1981, which told the story of two Aboriginal bands on the road in South Australia. The bands were Us Mob and No Fixed Address. No Fixed Address's first single, *We Have Survived*, was released at the same time as the film, and would go on to become, as the band's singer and songwriter Bart Willoughby describes it in *Road*, "a black anthem". Willoughby had been a student at the Centre for Aboriginal Studies in Music (CASM) based at Adelaide University. In an interview in *Road*, Willoughby describes what it was like growing up as a young Aboriginal man in Adelaide:

> I was a bad boy when I was young. When I say 'bad', I was in the game and I grew up on the streets, and so it all began when I was about nine right up until I was about 16, 17, and it was the toughest gang in Adelaide. And Adelaide is a tough town, so you had to be tough. It wasn't tough, it was just what I knew at the time, I didn't know I could play music. But you survive all that. Music and art somehow draws you out of there like a magnet.

For me, as co-producer and co-writer of *Wide Open Road*, those were the most important words in the entire series: "Music and art somehow draws you out of there like a magnet." They speak to a common experience of both indigenous and non-indigenous musicians and songwriters. But, once again, there's a depth of experience – a kind of authority in the grain of Bart Willoughby's voice, in the way he inhabits those words – which I would argue is utterly unique and distinctive to a radio history.[5]

Moreover, what the radio feature gives us is a conjunction of two things: not just an embodied cultural history, incarnate in the grain of the voice; but also a history of the *sound* of a cultural moment. It is not just Bart Willoughby's voice that we attend to; nor is it only the lyrics of *We Have Survived,* with its defiant challenge to white Australia and white Australian nationalism:

Cause we have survived, the white man's world,
and the horror and the torment of it all.
We have survived, the white man's world,
and you know, you can't change that.[6]

We also hear, unmistakeably, the sound of an important turning point in the development of Australian pop music; the first time Aboriginal musicians appropriated reggae as a musical form and made it their own. That moment is given an extra poignancy by the fact that it was a white musician, Graeme Isaac, director of CASM, and himself a film-maker and musician, who introduced Bart Willoughby to reggae, which would go on to become one of the defining sounds of Aboriginal rock music.

Kev Carmody, one of Australia's most prolific and respected indigenous songwriters, pays tribute to the significance of that moment in *Road*:

---

[5] A somewhat similar argument about the distinctive nature of radio as cultural history and the centrality of voice is developed by Siobhan McHugh. See McHugh, S. (2011). "Oral History and the Radio Feature: the aesthetic democratising of history". Paper delivered at Radio Studies Conference 2011: A Transnational Forum, Auckland University of Technology, Auckland, New Zealand, 11-14 January, 2011.

[6] *We Have Survived,* written by Bart Willoughby, appeared on the 1981 album *Wrong Side of the Road,* released by Black Australia Records and distributed by EMI Records Australia. All songs copyright US MOB and NO FIXED ADDRESS. I have been unable to find any other publishing details for this song.

It was just so uplifting and exhilarating to hear the accents, for a start, and
to hear language. Like No Fixed Address, Us Mob, it was just fantastic.

Once again, it is the centrality of voice that Carmody draws attention
to – hearing the accents of Aboriginal people singing in English. It is my
contention that in the conjunction of the grain of the voice and the sound
of the music, in the alchemy of the two heard and experienced
simultaneously, the radio feature embodies a unique history of sensibility;
an aural sensibility.

In making this claim of uniqueness, there is a danger of falling into a
kind of vulgar essentialism. As Lacey points out, radio researchers need to
be wary of "the trap of emphasizing radio's distinctiveness over its
similarities and connections with other cultural forms." Instead, she
argues, radio should be studied as "part of a wider matrix of
communications media." In this case, that wider matrix could and should
include the development of the local and international music industry, the
emergence of new marketing tools such as music videos and the rise of
new formats for presenting and publicizing pop music such as MTV.
Interwoven with these larger trends, however, is the specific nature of the
radio market in Australia from the mid-1970s onwards. In the preceding
part of this discussion, I have been concerned with radio as a way of
narrating cultural history. In the final section, I will now turn to the role of
radio itself in the making of that history.

One of the "back stories" which emerged from our interviewing for
Wide Open Road was the very important role which community radio
played in supporting and enabling Australian independent music in the late
1970s and early 1980s.

The early history of public radio in Australia, or community radio as it
has now come to be known, has been well-documented. In 1974, the
McLean Inquiry into public broadcasting had recommended the
introduction of FM broadcasting and the granting of public licenses
(Stafford, 2003, pp. 25-33). Shortly before the infamous dismissal of the
Whitlam Labor government by the governor-general in November 1975,
the government announced that 12 experimental licences would be given
to universities to set up campus stations. After some hesitation, the
incoming Liberal government, led by Malcolm Fraser, decided to support
the expansion of public radio. Over the next few years, more licenses were
granted to a wide range of community groups.

The founding of the first public radio stations such as 4ZZ-FM in Brisbane happened to coincide with the first stirrings of punk rock. In a chapter devoted to the early history of 4ZZ in his book *Pig City*, Andrew Stafford notes that punk was not immediately embraced by all of the station's staff or its listeners:

> It would be easy to jump to the conclusion that the mutually confrontational agendas of punk and Triple Zed were made for each other. The truth is its rise caused as much division within the station's ranks as it did almost anywhere else (Stafford, 2004, p. 39).

Nevertheless, punk quickly became a central if controversial part of the station's playlist, championed by presenter Michael Finucan (Stafford, 2004, p. 41).

In other cities, too, public radio stations were the first to play punk and new wave releases shunned by commercial AM or the new commercial FM stations. Many of these stations also had an explicit commitment to supporting and promoting Australian music and Australian songwriters – with a particular emphasis on artists from the cities where they were based. Put simply, public radio created a new audience for Australian rock music. It introduced a generation of musicians to a generation of listeners. According to Terry Bradford, program manager with Adelaide station 5MMM (now 3D-Radio) when it commenced broadcasting in 1979, "public radio was like an unexploded bomb" (Morton, 2008).

When that bomb went off, says Bradford, "...we were able to change the landscape of Australian music. I think public radio, was the battleground – for a musical reformation which was going on all the time "(Morton, 2008).

Michael Meadows, Susan Forde, Jacqui Ewart and Kerrie Foxwell have highlighted the role of community radio in creating an Australian community public sphere and facilitating new forms of cultural citizenship. As they argue, "local media [such as community radio], *both produce and maintain* the culture of a community" (Meadows et al, 2005, p. 172). However, neither they nor other academic researchers have focussed on the role of community radio in producing and maintaining the culture of Australian independent music from the late 1970s onwards.

Parts of the history of this symbiotic relationship between public radio and the independent music scene in the late 1970s and early 1980s have been documented by music journalists and writers working outside the academy. A more comprehensive history still remains to be written – or told on radio. Not only did community radio give airspace to musicians and songwriters who were ignored or rejected by mainstream radio; it also created a cultural space where listeners could recognize themselves and others like them. David Hendy has drawn attention to the particular role that music radio plays in creating imagined communities of listeners. "If it is true" says Hendy, "that through radio *we hear what we are*, it is also true that to some extent *we are what we hear*" (Hendy, 2000, p. 214).

Community radio stations such as 5UV and 5MMM in Adelaide (now 3AD), 4ZZZ in Brisbane, and 3RRR in Melbourne created a powerful sense of identity for the generation of music fans and listeners which came of age along with Australian popular culture.[7] As Helen Hambling, one of the founders of Triple Z, put it in an interview with Andrew Stafford, "what I think Triple Z did was it gave that subculture the capacity to communicate with itself" (Stafford, 2004, p. 36).

The situation in Australia could not have been more different from that in Britain, where punk and its many offshoots were excoriated in the tabloid press, but championed in print by magazines like the *New Musical Express* and *Sounds*, and on air by John Peel. The imagined community of listeners to whom Peel was broadcasting was a national, indeed an international community. In Australia, the new classic rock FM stations in Australia ignored bands like the Saints, Radio Birdman, the Go-Betweens, the Boys Next Door and their contemporaries. Yet these same Australian bands were exposed to a much larger British (and European) audience by Peel.

In this sense, the radio landscape in Australia was much more similar to that of the US, where it was college radio that nurtured what came to be known as independent music during the 1980s.

There was no national imagined community of listeners for Australian independent music until what was JJJ went national in 1989. That moment

---

[7] 2JJ, which 2JJJ and then simply JJJ, undoubtedly played a similar role in Sydney. But 2JJ was a special case; it was part of the Australian Broadcasting Corporation, and, as such, was not strictly speaking a public or community station, although it often behaved like one.

was celebrated as the creation of a national youth network with a distinct and distinctive musical and cultural identity. In my view, it was also the beginning of the end.

As I have argued above, one of the strong underlying themes of *Wide Open Road* is that the sound and the songwriting of bands like the Saints (from Brisbane) the Triffids (from Perth), and the Moodists (from Adelaide) grew out of a deep sense of isolation and alienation from mainstream Australian culture. For fans and followers of these bands and other like them, public radio created an oppositional taste-public quite distinct from the mainstream music culture of Oz rock, played out on commercial FM radio and in the suburban beer barns.

In the 1990s, the national audience which JJJ created was instrumental in the emergence of a whole new generation of Australian bands, again strongly influenced by American music; in this case what came to be known as grunge or the Seattle sound. Unlike their predecessors, bands such as Powderfinger, Spiderbait and Silverchair succeeded in attracting attention from major record labels and airplay on national radio, without needing to leave Australia to do so.[8] But this new generation did not continue the lyrical exploration of the landscape and Australian culture begun by the post-punk generation. That was left to a new and different musical subculture altogether—Australian rap and hip-hop—which again received (at least initially) very little airplay on either commercial radio or JJJ.

This transformation of the Australian radio landscape and the music industry was accompanied by, and imbricated with, the acceleration of globalization in the 1990s, and the whole spectrum of cultural and economic changes which went with it.

In *Coast*, the final episode of *Wide Open Road*, we, the writers and producers, chose to cast these changes in a broadly positive light. The story we told was broadly as follows: in the later 1990s and the early years of the 21st century, Aboriginal musicians and songwriters such as the Pigram Brothers and the Saltwater Band achieved prominence with songs rooted in the coastal landscape of Northern Australia, and reflecting both Aboriginal cultural tradition and the experiences of indigenous people working in industries such as pearl-diving. Their sound draws on a diverse

---

[8] See for example *The Sell-in: How the Music Business Seduced Alternative Rock.* (Mathiesen, 2000)

range of influences, from the surf guitar bands of the 1960s, such as The Shadows, through country music, to Pacific Islander musical traditions. The Pigram Brothers perform nationally and internationally, but continue to live in Broome, on the north-west coast of Western Australia, and have their own recording studio there.

In parallel with the rise of indigenous coastal music, a new kind of white Australian roots music begins to emerge, strongly connected with a coastal subculture in areas like the north coast of NSW and the south-west coast of Western Australia. This coastal culture is a hybrid of surf culture, the New Age and hippie culture of the 1970s and 1980s, and the slacker mentality of the grunge era. The music that grows out of it is also hybrid, an amalgam of folk, blues and alt-country influences which purports to be "rooted" in the coastal lifestyle and landscape, but is in fact, global (and in that sense, rootless) in its sound and lyrical pre-occupations. The coast becomes a meeting-point for the local and the global, and a metaphor for the porous and permeable nature of Australian culture.

In the closing words of *Coast*:

> From the Pigram Brothers in Broome to the Hilltop Hoods in Adelaide, Australian songwriters no longer have to leave our shores to find themselves, and to dream of the land they come from. We're more connected to the world than ever before, but it's the music of this place, this ancient country of dry brown plains and endless coastlines, of mangroves and eucalypts and dry brown scrub, that tells us something about who we are.

This narrative, which suggests a second coming-of-age of Australian popular culture, to use Clinton Walker's phrase once again, this time on the global stage, is one possible and plausible description of the trajectory of pop music from the mid-1970s to the mid-Noughties. However, I think it is not mere nostalgia which has led me to suggest above that the story we chose to tell in *Wide Open Road* is not the whole story, and that it conceals a counter-narrative of its own. In this counter-narrative, the intense period of musical and cultural creativity in Australia in the late 1970s and early 1980s grew out of local imagined communities in Adelaide, Brisbane, Perth and elsewhere, at odds with the dominant culture, and imagining an altenative culture and a relationship between that culture, their lived experience, and the landscape, which existed only in a fragmentary and provisory way. As I have suggested, community radio played an important role in that act of imagination. If, as Kate Lacey has argued, "the study of radio can draw attention to how, in effect, the sound has been turned down in much historical writing" (Lacey, 2008, p. 30), one might say that the sound of community radio was turned down in *Wide Open Road*. Turning it up again is, in my view, an important part of telling the cultural history of Australia in the late 20th century.

References

Primary Sources

Wide Open Road (2008). Retrieved April 14, 2011, from
http://www.abc.net.au/wideopenroad/

Unless otherwise indicated, all interview quotes in the text are from the audio of the four radio features available at the above web address. The title of the specific feature in which the interview occurs is given in the text.

Secondary Sources

Biddle, I. & Knights,V. (2007). *Music, national identity and the politics of location. Between the global and the local.* Aldershot, Hampshire: Ashgate.

Connell, J & Gibson, C. (2003). *Sound tracks. Popular music, identity and place.* London: Routledge.

Hendy, D. (2000). *Radio in the global age.* London: Polity Press.

Homan, S. & Mitchell, T. (2008). *Sounds of then, sounds of now. Popular music in Australia.* Hobart: ACYS

Madsen, V. (2009): A radio d'auteur: the documentaire de creation of Kaye Mortley. *SCAN. Journal of Media Arts Culture.* 6 (3). Retrieved April 14, 2011, from
http://scan.net.au/scan/journal/display.php?journal_id=142

Mathiesen, C (2000). *The sell-in: How the music business seduced alternative rock.* Crow's Nest, Sydney: Allen & Unwin

Meadows, M., Forde, S., Ewart, J., Foxwell, K.(2005). Creating an Australian community public sphere: the role of community radio. *The Radio Journal*, 3, (3), 171-187.

Mitchell, T. (2009). *Wide open road* – An audio roadmap of Australian popular music, *Music Forum* 15, (4), Retrieved April 14, 2011, from
http://www.musicforum.org.au/

Morton, T. (2008). Interview with Terry Bradford, Adelaide, 2008:

Unpublished audio recording.

Scannell, P. (2008). Sincerity. In A.Crisell (Ed.) *Radio*, Vol 1, (pp. 386-397), New York, NY: Routledge.

—. (2009). What is radio for? *The Radio Journal,* 7 (1), 89-96.

Stafford, A. (2004). *Pig city. From the Saints to Savage Garden.* St Lucia, Queensland: University of Queensland Press.

Tebbutt, J. (1989). Constructing broadcasting for the public, in H. Wilson (ed.) *Australian communications and the public sphere*, (pp.128-146), Melbourne: Macmillan.

Turner, G. (1994). *Making it national: Nationalism and Australian popular culture.* St Leonards, NSW: Allen & Unwin.

Walker, C. (1996). *Stranded: The secret history of Australian independent music 1977-1991.* Sydney: Macmillan.

—. (2000). *Buried country. The story of Aboriginal country music.* Annandale, NSW, Pluto.

CHAPTER SIX

"BOWIE'S WAIATA":
RADIO DOCUMENTARY AND FANDOM

SAM COLEY

On 22 April 1983, David Bowie's *Let's Dance* became the largest selling song on the New Zealand singles chart. The track stayed in the number one position for the following six weeks, until it was eventually replaced by Michael Jackson's *Beat It. Let's Dance* was Bowie's first single and the third track off the album of the same name, which produced the subsequent hit singles; *China Girl* and *Modern Love*. The release of *Let's Dance* was New Zealand's first encounter with what was to become the most commercially successful period of Bowie's career to date. Although many critics have claimed that the album marked the beginning of a decline in Bowie's creativity, its international popularity in early 1983 paved the way for the massive world tour that followed.

The show, called the "Serious Moonlight" tour, commenced in Belgium in May 1983 before travelling across Europe and America and then down to Australasia, reaching New Zealand in November of that year.

Shortly before his first New Zealand concert at Athletic Stadium, Wellington on 24 November 1983, Bowie was invited by the indigenous Maori tribe "Ngati Toa" to visit Takapuwahia marae in nearby Porirua[1]. The invitation was a rare honour at the time and saw Bowie become the first rock star to be officially welcomed onto a Maori marae. Remarkably, Bowie composed and sang an original song for the people of Ngati Toa that was captured by a radio station that was also present. The song was

---

[1] The marae is a sacred place in Maori society, used for important communal events such as religious and social ceremonies and the welcoming of important visitors.

never officially released and was only heard on the marae that day in 1983 and in a radio news report on Wellington's 2ZB the following morning. It was then kept in storage at a New Zealand radio station. Twenty-five years later, this event and the song Bowie sang became the focal point for a series of radio documentaries and audio slideshows designed to commemorate the occasion.

This article will discuss the production of these documentaries and other related content which I shall refer to as the "Bowie project", along with the activities of the Bowie fan community before and after the broadcast and web-streaming of the documentaries by Radio New Zealand and Radio Hauraki. At this point, I must declare my bias as being a paid-up member of the David Bowie Fan Club. Although I was too young at the time to personally attend the *Serious Moonlight* tour, my older brother did, and I can clearly recall the nationwide "Bowie Mania" that swept the country at the time. As a radio documentary producer, the twenty-fifth anniversary of the tour provided me with an opportunity to capitalise on my own personal fandom by creating original content recounting the story of Bowie's visit to Porirua and the record-breaking concert at Western Springs in Auckland that followed.

The subsequent documentaries I produced, *Bowie's Waiata* and *Down Under the Moonlight*, are central to this chapter. I shall detail the approach taken to the production of these documentaries and define the attributes of radio as a medium for music documentary. Another critical aspect of the Bowie project is how the material was remixed and re-appropriated by Bowie fans. I shall explore the notions of audience and fandom towards the project and demonstrate some of the potentials of digital content creation, sharing and networking for radio work. This shall encompass the role that user-generated content provider YouTube played in continuation of the story after the initial radio broadcasts of the documentaries. It is both a story of the re-appropriation of a moment in time and of the re-appropriation of that project by its audience.

Online promotion

The Australasian fan website *Bowie Down Under*[2] endorsed and promoted the *Bowie's Waiata* documentary prior to the Radio New Zealand broadcast on the 22nd of November, 2008, alerting both the

---

[2] www.bowiedownunder.com.

Australasian and global Bowie fan community. In the days following the original broadcast, a member of the Bowie fan community uploaded an unauthorised recording of the documentary to a file-sharing website. Several different fan edits of the documentary using sections of the original content appeared on *YouTube*, providing their own interpretations of the *Bowie's Waiata* documentary. This chapter examines the activities of the David Bowie fan community and includes examples of how fans engaged with the documentary. This demonstrates how listeners were able to become producers themselves and continue the story of the documentary through their own examples of media production.

The *Bowie's Waiata* documentary focused on Bowie's visit to Takapuwahia marae and predominantly featured interviews with members of the Ngati Toa tribe. The *Down Under the Moonlight* documentary was centred on the historic concert that took place at Western Springs Stadium, Auckland, on the 26th of November 1983.The remarkable success of the *Serious Moonlight* tour in New Zealand was not entirely surprising as he had an extensive fanbase in New Zealand. Bowie's previous tour of Australasia was the *Low / Heroes* concerts of 1978 which played across New Zealand and Australia. These concerts were critically acclaimed and judged to have been an enormous success. This was particularly impressive considering that Bowie was not promoting a "hit" album at the time. However, as these shows were New Zealand's first opportunity to see Bowie live, immense public interest was assured. According to the website *Bowie Down Under* the Auckland concert at Western Springs Stadium on the 2nd of December 1978 set a new national attendance record. At the time, the audience of 41,000 was considered the largest in New Zealand's history. However this record was broken, once again by Bowie, when he returned five years later with the *Serious Moonlight* tour.

According to The Herald newspaper, the Western Springs audience for this concert was estimated to be 80,000 (1983) - and was deemed to be the country's "largest rock extravaganza" in The Sunday News (1983). This represented the biggest single crowd gathering in New Zealand and was credited in the Guinness Book of World Records as "the largest crowd gathering per head of population anywhere in the world" (McWhirter, 1984). An advert taken out by Bowie promoter Paul Dainty in the US music industry magazine Billboard (1983) claimed that the size of the Auckland concert was enough to make it New Zealand's fifth largest city in terms of population. However, an online Bowie fan site called the official figures conservative, given that the perimeter fence of the stadium was pulled down during the concert, allowing thousands of non-paying fans unofficial entry to the event.[3]

The *Bowie Down Under* website claims:

> The national attendance records set particularly in New Zealand made this tour relevant in any reading of the nation's modern cultural history. In terms of Bowie's career and the perspective of rock music, it is a testament as to how big a cult artist can become.

The cultural significance of the visit, coupled with the widespread popularity of Bowie's *Serious Moonlight* tour amongst New Zealanders old enough to remember it led to the author being commissioned to produce two original radio documentaries to play in November 2008 to celebrate the twenty-fifth anniversary of the tour.

The content of *Bowie's Waiata*

In 2001 I found an audio file of Bowie singing live at Takapuwahia marae in 1983 in the archives of The Radio Network, a commercial broadcaster in New Zealand. The audio was a previously unheard track, "Waiata" (Maori for "song") that had been recorded during Bowie's visit to the marae. It is protocol for guests being welcomed onto a Maori marae to respond to the sung waiata of the tribe by singing a song in return. However, most visitors do not write an original song to mark the occasion as Bowie did. He was clearly made aware of the importance of the welcoming ceremony and can be seen to have responded with a degree of sincerity. Although this song was apparently recorded in less than ideal conditions and approximately 30 seconds in length, its rareness as the only known recording of an original Bowie composition made it a useful foundation for the project. The track was digitally enhanced through compression, graphic equalization and volume enhancement, then looped in the final broadcast to extend its length. Bowie backing singer Frank Simms was played the track at a recording studio in New York and his immediate response to hearing the track for the first time in twenty-five years became a highlight of the final *Bowie's Waiata* documentary. Simms had no idea that the performance had been recorded and had last heard the song while performing it live, acappella style, with his brother George

---

[3] "Numbers had swelled considerably by at least 20,000 due to people pushing the perimeter fences over." Comment from "Paul" www.bowiedownunder.com.

Simms and Bowie at Takapuwahia marae (Simms, March 10th 2008). Further information regarding the source or context of the audio was not available, as it was simply labelled "Bowie Porirua Marae 1983". However, as the song had not been listed as an official Bowe release and had no apparent publisher, the audio was deemed not to infringe copyright by the relevant broadcasters and formed the starting point for two radio documentaries, one online version, eight *YouTube* clips and four on-demand web "featurettes".

The initial documentary, *Bowie's Waiata* was produced for the public service broadcaster Radio New Zealand, while the second documentary *Down Under the Moonlight* was produced for a commercial rock station, Radio Hauraki. As these radio stations have very different identifies and audiences both documentaries required a separate approach to production. The former underplayed Bowie's "hits" in favour of a more subdued sound, utilising traditional Maori instrumentation as backing beds. In contrast, the Radio Hauraki documentary was structured to fit around a commercial schedule and included many of Bowie's best-known songs with an emphasis on tracks already featured in the Radio Hauraki playlist. This mimicry of the "station sound" assisted the documentary to blend in with regular programming, which did not traditionally feature documentaries.

Music documentary

When considering the field of music related documentaries it is useful to consider the wider attributes of radio as a broadcast medium. In this section I will define the principle similarities and differences that exist between radio and visual forms of music documentaries. I shall explore these comparisons along with the role of the radio documentary producer and the techniques they employ. The production of television and film documentaries has been well represented in many academic publications. However, there are relatively few on the subject of radio documentary production. It is therefore not surprising that academics have tended to favour the moving image over audio (Lewis, 2000; Tacchi, 2000).

The increasing availability of digital media and the improving fidelity and affordability of home audio-visual systems has assisted the popularity of the music documentary as a genre. Specialist music television channels and formats such as DVD and Blu Ray have allowed fans to enjoy a variety of high quality visual documentaries while the internet has made both new and archived audio documentaries more easily accessible.

There are many comparable aspects to these two forms of music documentary production. In his book *Speech, Music and Sound* Theo van Leeuwen (1999), comments on some of these parallels by observing that sound dubbing technicians in the radio, film and television industries follow similar approaches to the categorisation of sound tracks by separating them into three spatial zones, close, middle and far distance.

Sarah Sherman's article *Real(ly) Good Stories* (2010) reported on discussions held during *Doc NYC 2010*, a festival celebrating the documentary form. Sherman noted comparisons between the structure and concepts of both radio and film documentary, which achieved "a singular texture and strength". Sherman also notes that various film documentary production techniques can also be found in radio documentary production. An example of this is the use of stock footage (or audio), interview segments, narration and scene recording.

However, when identifying differences between these two forms of media the radio producer can often been seen to enjoy a far greater level of creative input than their television counterpart. David Hendy (2004) describes radio production as a less technically complex process than television, which gives radio producers a much greater degree of creative freedom than producers involved in television and film documentaries. Hendy suggests that radio is more of a producer's medium than television since various roles such as researcher, director, editor, sound recordist and presenter can be combined into one role – that of a multi-skilled radio producer. McLeish (1999) claims this multi-skilling is part of a production "convergence" that has led to a reduction in production costs, which has allowed more radio programmes to be made by fewer people. This increased control has allowed the radio producer to exercise a considerable amount of freedom as their production decisions do not require the approval of the co-producers and technicians often required for television and film production work.

The primary difference between these two forms of documentary production is of course radio's absence of visual information, which led to Andrew Crisell (1992) referring to it as a "blind medium". The listener is required to create their own "pictures" which, as Gulson Kurubacak (2004) contends, gives them the ability to control and extract their own meanings in their minds. This creative participation from the listener, when combined with music's ability to trigger emotion, can create a very personal response to a production, which can build substantial trust between a documentary and its audience. In the preface to Jeanette Bicknell's book *Why Music Moves Us* (2009), listeners have recounted being "overwhelmed or overpowered by music, reduced to tears, and experiencing chills or shivers and other bodily sensations." By exploiting the emotion a fan has towards elements of a favourite song, such as instrumental sections, choruses and bridge sections, the radio producer strategically manages the listener's interest in order to propel the narrative forward. The producer, by demonstrating their own passion for the subject, establishes a connection with the listener, earning their appreciation of a production's worth.

Given that this chapter explores the relationship between fandom and radio documentary production, it is worth considering how an audience can have a profound impact on the initial stages of the production process. Guy Starkey (2007) sees the radio documentary as a means of communicating a story, which the producer expects an audience to be receptive too. The influence of perceived fan expectations can therefore be seen as being fundamental to the production process. David Hendy (2004) suggests that the selection of an appropriate approach, which takes into consideration the context and audience of a documentary can be of equal importance to the proposal stage of a project as the actual choice of subject and content.

Consuming the documentary online

Although the initial Bowie documentaries were consumed live and simultaneous by a traditional radio audience via terrestrial broadcasting, in their online "on-demand" form they accumulated individual listeners in the form of "hits" therefore incrementally building a continuous audience throughout the duration their existence on the Internet. This section shall examine how David Bowie fans responded to the *Bowie's Waiata* and *Down Under the Moonlight* documentaries online. Before the advent of the Internet, fans would often be alerted to upcoming radio documentaries through promotional commercials and by way of announcers talking about an upcoming item on the broadcast schedule or through publications such as the United Kingdom's *Radio Times* and the radio listings section of newspapers. The development of online chat-rooms, dedicated fan-sites, *Facebook* and RSS feeds has offered fans highly detailed information about upcoming programming specific to their particular interests and provides easy access to previously broadcast radio documentaries which have been unofficially archived by fans.

The online promotion of *Bowie's Waiata* began when a member of the website *Bowie Down Under* saw the documentary listed in the New Zealand publication *The Listener* and alerted the fan community to its upcoming broadcast. On noticing this activity, the website was contacted to promote the *Bowie's Waiata* documentary and *Down Under the Moonlight* to a target audience of fans who had already demonstrated their interest in Bowie's Australasian activities. The *Bowie Down Under* site identifies itself as "the David Bowie community of Australia and New Zealand" and was a rich source of pre-production information and photographic content for inclusion in the accompanying audio slideshows. Adam Dean, the webmaster of *Bowie Down Under* was clearly enthusiastic about the project, posting a number of alerts on the site. These detailed programme content and musical track listings alongside new photographs taken during the process of interviewing Bowie's associates from 1983, often holding items of memorabilia from the *Let's Dance* period.

Promotional postings provided an international audience of Bowie fans with specific broadcast information, such as links to the Radio New Zealand and Radio Hauraki websites where the documentaries could be streamed live. The site also enabled fans to convert New Zealand's standard time zone into that their own country in order to hear the documentary streaming at the correct time. Other Bowie fan websites picked up on the *Bowie Down Under* story and their chat-rooms and forums featured several postings in which fans questioned certain aspects of Maori culture, such as the meaning of the word "waiata".

These online communities provided myself as the producer an opportunity to interact with the fan audience and respond to questions about the production. There was also fan activity noted on the forums of the official David Bowie website[4], although, as an unsanctioned documentary the project was not officially referred to on this site.

The majority of discussion relating to the documentary found on the official Bowie site revolved around the inclusion of the unreleased Bowie song *Waiata*. By posting in these online fora, I was able to build a relationship with the fans and in turn rely on them to promote the documentary to a global audience rather than just its physical broadcast location of New Zealand. Unexpectedly, rather than simply promoting the documentary, a certain section of the online David Bowie fan community captured the audio from the broadcast and made it available online for those outside of New Zealand to experience.

---

[4] www.davidbowie.com.

In an effort to combat copyright infringement, Radio New Zealand asks that producers provide two separate mixes of their commissioned documentaries: one that contains copyright music and another that has all copyright music removed. Although the station has a license to play documentaries with music content in broadcasts, this does not cover the "listen again" content on their website. By making a music-free version available online, fans still have access to the bulk of the spoken word content. This avoids potential legal issues with record companies and is supposed to stop listeners illegally capturing content paid for and owned by Radio New Zealand. However, this practice clearly did not work in the instance of the Bowie project. Several audio files of the full documentary were found to have been captured by Bowie fans from the live webstream of *Bowie's Waiata* on Radio New Zealand's website. These were then uploaded to the internet for listeners outside of New Zealand to hear. I found the documentary available on the direct download site *4shared* and it was also hosted directly on the *Bowie Down Under* website.

The capturing of audio broadcasts from a radio station is clearly not a recent practice. Radio listeners have been recording broadcasts since the audio cassette format made convenient analogue recording common from the mid-seventies. However, as many stations stream their live digital content online, international audiences are now able to capture international audio, instead of just local radio content, through the use of streaming audio capturing software such as *Totalrecorder*, *StreamripperX* and *Audio Hijack*. Some of these digital applications have allowed fans to record a direct digital signal with no discernable loss of audio quality. Users can then save these recorded files as standard .wav or .mp3 formats. These files can then be easily uploaded to online file sharing sites or to web content providers, as was the case in the Bowie project.

Following the broadcast of the documentaries, several examples of fans gathering and reusing the content to create new media artefacts were identified. This fan practice can be considered as a form of remixing or re-appropriation as the original content is reinterpreted and reshaped to create new content. In his book "Remix", Lawrence Lessig (2008) interviewed remixer Gregg Gillies who described this practice as being positive since it was in effect a "free culture". Gillies observes how ideas impact on data by allowing manipulation, treatment and ultimately further dissemination.

Although some fans carried out remixes of the original documentary by capturing audio from the Radio New Zealand webstream, other fans resorted to more creative means of obtaining content for their productions. Once the initial documentary had been broadcast, several emails were received, offering feedback and requesting further information. An email from a listener using the name "Kristi" provided mild flattery and a nostalgic back-story to assist her request for a CD version of the documentary:

> I look forward to hearing the Down Under documentary. I heard your Waiata one and thought it was very well made and just fascinating to hear about, especially as I was a little 7 year old who lived up the road from the marae he visited in Porirua. I ran down to have a look at this Bowie guy that everyone was going on about. Thanks again. ("Kristi", email, 3 February 2009.)

As the CD was assumed to be heard only by an individual listener, this request was deemed not to unduly infringe on copyright regulations. Therefore a copy was sent to the address given and nothing more was heard. However, approximately one month later a series of three *YouTube* uploads were noticed. These took the form of three audio slideshows using related archival photographs, apparently sourced from the Internet, combined with high quality audio from the *Bowie's Waiata* documentary. The producer of these unofficial features had listed myself as the original writer and producer along with Radio New Zealand as the broadcaster and accompanied these credits with the date of production and, interestingly, the international copyright symbol. The username for these audio slideshows was credited as "Bodacea1". On contacting "Bodacea1" via YouTube messaging it was revealed that the producer was in fact "Kristi". Another *YouTube* user "MrDavidBowie" is credited as the producer of another audio slideshow based on content sourced from the original *Bowie's Waiata* documentary. However, this version used inferior audio, which is assumed to have been captured from Radio New Zealand's webstream of the documentary. In the examples provided by "Bodacea1" and "MrDavidBowie", both had re-appropriated the original audio by selecting new edit points, dividing the audio into chapters and adding their own accompanying visual elements. This activity reflects the observations of Stefan Sonvilla-Weiss (2010) who refers to the unfinished nature of digital artefacts, which he contends continually evolve through an ongoing process of development.

Following the presentations of this research different academic fora[5] audience members have questioned my response to having work remixed by Bowie fans without specific permission and often un-credited. In regards to the case study provided by the Bowie project, I view the re-appropriation of my work as a constructive practice that has added new interpretations and greater depth to the initial story. In fact, the original Bowie documentaries contained elements of re-appropriated audio taken from previously produced Bowie documentaries. Fan-generated remixes of the story can be seen as almost inevitable by-products of online media distribution and consumption. As these fan-generated productions have not been financially motivated and have only generated relatively small amounts of online activity, no known legal action has been taken (to date).

The re-appropriation of content from the Bowie project by the fan community was not the only way the story was added to and enriched. Shortly after the documentary was broadcast, an Auckland based journalist, Greg Ward, contacted Radio New Zealand who in turn forwarded his email to me. This communication revealed him to be the original sound recorder of the *Bowie's Waiata* song at Takapuwahia marae in1983. Ward kindly agreed to write an online article detailing the story behind his capturing of the audio. This additional text content is currently accessible alongside audio and pictorial images from the documentaries and, in a sense, can be seen to continue the story by providing added depth and clarity to the narrative. Ward takes up the story:

> On the day of the marae visit I was a young reporter working for Radio
> New Zealand News. I was probably covering the event in my own time.
> Being a huge Bowie fan, I had taken at least several days leave to focus
> entirely on a getting an interview during his visit. (Gregg Ward, email 10
> December 2008)

Ward goes on to provide an explanation for the poor quality of the recording:

---

[5] Previous versions of this chapter have appeared as conference papers at "Internet Attractions: Online Video and User-Generated Ephemera", University of Nottingham, June 2009, the "Sights and Sounds Conference; Interrogating the Music Documentary" University of Salford, June, 2010, and "The Radio Conference: A Transnational Forum", Auckland University of Technology, 11-14 January 2011.

An external PA system suddenly fired up and the waiting crowd began to follow proceedings inside the marae, via loudspeaker. I'm sure Bowie and his management were completely unaware we could hear everything outside. And when it was announced that Bowie was going to sing, I made sure my $50 microphone was pointing in the right direction – up close and right beside a loudspeaker on a perimeter fence. This must have been David Bowie's most unsophisticated recording ever. Talk about unplugged! (ibid.)

The role of You Tube

Many radio productions are available in edited or full form, often in a series of clips via user-generated content providers such as *YouTube*. This audio includes captured radio programmes, podcasts, bootlegged concert broadcasts, live sessions and interviews. In January 2009, 100.9 million United States-based Internet viewers watched 6.3 billion videos on YouTube according to data from ComScore Video Metrix (2009). Page views and video views from this large online audience provide useful indicators that can inform producers about the consumption of their online content. However, in the early days of the internet, this data was often limited in terms of specific detail. To rectify this, *YouTube* launched a feature called "Insight" in March 2008, a video analytics tool that offers users specific information on user engagement with their content. Features include where a video's viewers are geographically located, which parts of the video command the most attention and the gender and age range of the viewer. This statistical data can provide the producer with an accurate audience profile, an insight into how they discovered the item and which parts were most interest. "Community engagement" can also be measured by user comments, ratings, and "favourites" functions.

The Bowie project utilised *YouTube* as a means of distributing specific sections of the original documentaries along with extra (bonus) content that had not been used in the final edit. By creating a series of individual clips, groups of fans are encouraged to engage with content they might not have otherwise found. This is assisted by the "recommended for you" function in *YouTube*, which allows viewers to find related content easily. An example of this is the section of the *Down Under the Moonlight* documentary relating to the late Blues guitarist Stevie Ray Vaughan. Although Vaughan played on most of the *Let's Dance* album and was booked to take part in the *Serious Moonlight* world tour, he was replaced at the last minute in circumstances that were unclear. An interview with backing singer Frank Simms recounted his memories of events leading up

Vaughan being controversially "fired" from the band. A specific *YouTube* clip focused solely on Simms recollections of Vaughan's departure and attracted the attention of both Bowie fans and fans of the Blues. This collision of fandom sparked several aggressive exchanges in the comments section of the clip with some being automatically removed by *YouTube*, accompanied by the statement *"This comment has received too many negative votes".* [6]

Many comments questioned Simms' memory of events, which may have prompted the *Serious Moonlight* saxophone player Stan Harrison to post the following comment:

> Yo Frank. Stevie later told me that he seriously considered smashing David with his guitar - glad he didn't. I'd never seen him so upset. I do believe that Stevie was left standing at the door of the bus with his guitar and luggage because he didn't sign the contract, holding out for more money, as you said. (Stan Harris, *YouTube* comment, 15 September 2010)

This version of the story was contradicted by Craig Hopkins, author of the book "The Essential Stevie Ray Vaughan", who contacted me after seeing the *YouTube* clip:

> I interviewed Carlos Alomar, the band director, and Carmine Rojas for my biography of Stevie. COMPLETELY different story. This Simms guy is the only person from either camp who claims Stevie was fired. He wasn't. (Simms) got his facts wrong on just about everything else. Everyone sees a little slice and reaches their own conclusions. (Craig Hopkins, email 20 October 2010)

It is interesting to note the involvement of actual band members as well as experts, in the *YouTube* comments sections. This interactivity between producer, fan, expert and artist provides a valuable forum for discussion and to add nuances to a story, as well as the opportunity to verify the reliability of content.

---

[6] See
http://www.youtube.com/watch?v=2UdThAKKX8I&feature=channel_video_title.

Conclusion

Bowie's *Serious Moonlight* tour remains his longest and most successful world tour. Immediately after the New Zealand leg, Bowie and his entourage travelled across Asia, ending the tour in the Hong Kong Coliseum on the 8th of December 1983. In total, Bowie visited 16 countries, performed 96 shows and sold 2,601,196 tickets (Flippo, 1984). He returned to Western Springs four years later with the *Glass Spider* tour, although the show was not as well received as the *Serious Moonlight* tour, with reviewer Cameron Officer calling it a "disappointment" (2004).

David Bowie has not released an album since *Reality* in 2003. He underwent cardiac surgery towards the end of a comprehensive world tour in 2004 and has made few public appearances since. He has not toured or released any new material and has only made rare live appearances as a guest performer. Interviews have been scarce, although retrospective box sets from his back catalogue, along with comprehensive biographies and coffee-table books, have all helped to preserve Bowie's status amongst his fans. The lack of new material from Bowie has, in a sense, provoked many fans into cannibalising Bowie's earlier work in order to practice their fandom. The reworking of his back catalogue has taken many forms such as *YouTube* clips of songs with new images added, collections of online archives containing rare tracks and live recordings available through file sharing sites, and "mash-ups" where fans use digital editors to combine several songs into one new track. Bowie himself has sanctioned the official release of certain popular "mashups" and has approved the release of two iPhone applications, which allow users to remix separate instruments from well-known songs through a basic touch-screen mixer. In the example provided by *Bowie's Waiata*, fans have taken matters into their own hands by producing original content that is primarily designed to be consumed within fan communities. Although these artefacts seldom attract mainstream attention, the limited nature of their audience consequently allows fans to often avoid copyright infringement litigation.

This chapter has explored the ways in which David Bowie fans responded to radio productions based on events surrounding his New Zealand tour of 1983. I have identified that both prior to and following the broadcast of these documentaries fans were engaged in the practices of promotion, archiving and remixing of radio work that provided them with new material to further enhance and extend their connection with the artist and other fans. These practices would seem to indicate that the boundaries between producer and fan are becoming increasingly blurred. As the producer of this documentary, and a serious Bowie fan myself, I was able to draw on my fan knowledge to inform certain production decisions. Subsequently, some Bowie fans that consumed the documentary became producers of media by constructing and sharing their own interpretations of the documentary. By having fans re-appropriate the documentary and add their own interpretations the story presented in the original documentary does not end; it continues. Some of the re-appropriations of *Bowie's Waiata* discussed here added images to the audio, giving further context to the audio. Fans have also edited out pieces of the original documentary, making sophisticated editorial decisions about audio that is of particular relevance to them.

Although these fan produced items cannot be viewed as true collaborations, since they were created independently, these activities provide evidence of a new form of participatory culture where fans make use of new media technology to re-imagine media content. I argue here that this re-appropriation of material by fan communities is a constructive practice that ultimately enhances the production of contemporary radio documentaries by allowing the story to continue beyond the original piece, giving more context to the material and by providing a valuable forum for producers to interact with their audiences.

References

Bicknell, J. (2009). *Why music moves us*. London: Palgrave Macmallian.

Billboard Advert, Bowie Makes History Down Under. (10 December 1983). *Billboard magazine*, p.11

Crisell, A. (1992). *Understanding radio*. London: Routledge.

Flippo, C. (1984). *David Bowie's Serious Moonlight, the world tour*. New York: Dolphin Books.

Hendy, D. (2004). *Radio in the global age*. Cambridge: Polity Press.

Kurubacak, G. (2004). The building of knowledge networks with interactive radio programmes in distance education systems, Distance Education Department, College of Open Education, Eskisehir Anadolu University. Paper presented to *E-Learn World Conference*, Washington DC. Retrieved May 15, 2011, from http://www.editlib.org/p/11233

Lewis, P.M. (2000) Private passion, public neglect: The cultural status of radio. *International Journal of Cultural Studies*. 3 (2).160-167.

Low / Heroes Tour. Retrieved March 4, 2011, from

http://www.bowiedownunder.com/lowheroes/11.html

Officer, C. (2004). David Bowie blows away Wellington. Retrieved August 23, 2011, from http://www.thread.co.nz/news/760/15/David-Bowie-Blows-away-Wellington

Major Bowie controls 80,000. (28 November 1983). *The Herald*, p.3

McLeish, R (1999). *Radio production*. Oxford: Focal Press.

McWhirter, N. (1984). *The Guinness book of records*. New York: Bantam Books.

Sherman, S. (2010). Real(ly) good stories. Retrieved October 11, 2011, from http://www.mediarights.org/news/really_good_stories

Sonvilla-Weiss, S. (2010). *Mashup cultures*. New York: Springer-Verlag/Wien

Starkey, G. (2007). *Radio in context*. London: Palgrave Macmallian.

Tacchi, J. (2000). The need for radio theory in the digital age. *International Journal of Cultural Studies*. 3, 2, pp. 289-298.

van Leewen, T. (1999). *Speech, music, sound.* London: Palgrave Macmillan.

Violence rocks concert. (27 November 1983). *The Sunday News*, p.2

YouTube surpasses 100 million U.S. viewers for the first time. Retrieved June 2, 2011, from http://www.reuters.com/article/2009/03/04/idUS230743+04-Mar-2009+PRN20090304

Chapter Seven

The Centralisation of Regional Radio: City Versus Country in the Super Radio Network

Harry Criticos

Introduction

In 1992, the then Keating Labour government in Australia introduced sweeping changes to the *Broadcast Services Act 1942* (BSA) effectively deregulating the radio industry. Rather than delivering competition and diversity, the deregulation of radio resulted in a perceived loss of localism, a concentration of ownership, a lack of diversity and an increase in networking. One consequence of networking was the loss of local voices. There are now fewer journalists and announcers working in regional radio, many stations operating with a single announcer, and many stations without a journalist.

Networking is not a new phenomenon in Australian radio, but its increasing presence is a cause for concern in how regional radio functions in the licence area to which it broadcasts. While much has been written about networking and the changes to local content brought about by the BSA, much of the focus has been on the impact it has had on journalism, the delivery of news services and the effect on the diversity of information available to regional areas.

This chapter looks at the Super Radio Network (SRN) which operates 38 radio stations: 36 analogue and two digital services. The SRN has taken advantage of the lifting of ownership limits by networking an average eighteen hours of programme per day across its regional AM and FM stations. The research presented here indicates that programme makers in regional markets are at odds with those at the networking "hubs". It can be seen that programme makers at the hubs consider networking and localism as inconsequential, and put minimal effort into programming and programme content that is delivered to regional markets. Through interviews with management, announcers and journalists at an SRN hub and a regional SRN station, this chapter will look at the tensions between networking hubs and their associated regional stations.

The SRN is an example of how a company has derived benefit from a deregulated radio industry and has monopolised many of the licence areas in which it operates. It is evident that as the network expands, there is also an increase in the number of networked programme hours. The industry argues that networking is providing a service to a licence area that may not otherwise exist. However, data collected from interviews and observations at an SRN networking hub shows that the quality of the programme content fed to the regional network may be an issue. There seems to be little, if any, communication on the content of programme broadcast to the regional station by the hub. This, for example, can result in generic programming with no opportunity for the regional station to include local issues within that networked programme to cater for their licence area. There also seems to be a divide between regional stations and the feeder stations, creating an "us versus them" situation. This chapter offers evidence from the case study of the SRN as an example of a network that appears to operate at the bare minimum of the legislated requirements in offering a service to regional areas in New South Wales and south-east Queensland, but also discusses how local programme-makers attempt to maintain a level of local content alongside the city centric programming provided by the hub.

The Super Radio Network

The Super Radio Network is a subsidiary of the privately owned company Broadcast Operations Pty Ltd. At the time of writing it consisted of 39 radio stations comprising 22 AM, 15 FM radio stations and two digital stations. This makes the SRN the second largest radio network in Australia and the largest in NSW. The headquarters of the network is in Sydney. Located here are 2SM and the two digital stations. The other stations in the network are spread across northern New South Wales and south-east Queensland. The SRN runs two distinct formats throughout its network: music on the FM stations and talk on the AM stations. While there are stations that broadcast locally from 5:30 a.m. to 6 p.m., the majority of the regional SRN stations take networked programme from 9 a.m. till 5:30 a.m. the following day. The distribution of this networked programme is not limited to one central feeder station. Rather, it is divided between stations in Sydney, Newcastle and Tweed Heads, all located on the New South Wales coast. These stations share the responsibility of supplying programme content to stations owned by Broadcast Operations Pty Ltd.

Deregulation: an overview

The radio industry has undergone many changes to adapt to its operating environment. These changes have included a reaction to the introduction of television in the 1950s, through to dealing with new media such as the Internet, as well as acting on legislative changes that either increased or decreased the operating conditions of a radio broadcaster's licence. Following the enforced change of content that the introduction of television imposed on radio, radio has endeared itself to the listener by being able to relay to that listener both the future and the past by means of the news which broadcasts current events, competitions which may involve historical events, the announcer who talks of the past, present and future which allows radio to become a medium that is intimate, immediate and relatable (Potts, 1989).

Radio, worldwide, was a highly regulated industry until recently when governments generally chose to loosen regulatory controls in an effort to increase diversity (McQuail, 2005, p. 34, p. 239). In Australia, the industry is currently regulated by the *Broadcast Service Act 1992*. To understand the impact the process of deregulation has had on Australian radio over the last two decades, we need to understand the uniqueness of the Australian radio industry before deregulation.

Australia is a large country, where many people in rural areas live in isolation. For this group local radio is a means to gather information about what is happening in town; it has become a companion to many people (Griffin-Foley, 2004, p. 541). According to Lesley Johnson (1981) the Australian broadcasting system was unlike that of the US where there was no publicly owned broadcaster, or the UK where the BBC was totally state controlled. In Australia, prior to the introduction of community radio licences and other forms of broadcasting such as Internet streaming, radio operated under a dual system of broadcasting that Johnson states as having:

... facilitated a context in which competition produced vitality, while also providing an alternate national network to ensure that the public service aspects of broadcasting were fulfilled. In this manner the dual system was represented as a testimonial to the inherent value of competition, to the laissez-faire principle, while retaining the benefits of a watchful, though largely non-interfering, state. (1981, p. 173)

During this period, media regulation in Australia concerned itself with licensing, and operational and technical standards (Jolly, 2007, p. 63). From the 1930s successive governments introduced media rules ranging from ownership controls to local content regulations (Jolly, 2007, pp. 64-5). The changes to the *Broadcast Services Act* in 1987, and again in 1992 for radio broadcasting, reflected the changing economic philosophy of the Hawke and Keating Labour governments which regarded deregulation as a panacea for the nation's well being (NMA, 2008). To a certain degree, these changes to the way radio would function in Australia brought it into line with radio regulations of other countries. Changes were made to local content, foreign investment and ownership limits, and were designed to introduce cross-media ownership provisions that effectively deregulated the radio industry (Collingwood, 1999, p. 12).

The body that represents and lobbies for the commercial radio industry, Federation of Australian Radio Broadcasters (now Commercial Radio Australia), regarded deregulation as a way to "improve efficiency and facilitate expansion" (FARB, 2000, p. 10), a position that contrasts markedly with the interventionist approach of the BSA 1942. Hendy (2000, p. 43) asserts the industry's position by suggesting that deregulation has not gone far enough and that the "micro-management" by government (the regulation of content) should be relinquished so a radio station can meet the demands of its audience unhampered.

Despite the radio industry's desire for a further deregulated environment, and an increase in the number of licences issued by ACMA between 1992 and 2004 (Ames, 2005; Fairchild, 1999; Marcato, 2005), the industry has experienced a concentration of ownership rather than the competitive environment that was to be brought about by deregulation. This situation has created a problem for regulators because, as Braman (2007) states, media concentration through deregulation has "reduced the number of points at which pressure needs to be put by those who might be interested in shaping content" (p. 275). With fewer owners holding a large number of regional radio licences and networking a large number of programmes, the lack of a competitive environment makes it difficult for ACMA, for example, to apply tighter local content regulations on owners.

The development of networking

Networking, according to Shingler and Wieringa (1998), involves:

A station that broadcasts nationally or a large broadcasting operation on a regional or national level characterised by links between individual radio stations capable of sharing source material and output. (p. 5)

Since the deregulation of the Australian radio industry, brought about by the introduction of the *BSA 1992*, there has been a dramatic increase in the number of stations that take programmes from a central feeder station. This is very different to the regulated radio landscape that existed prior to 1992 in which, according to Armstrong, there existed a high degree of diversity through non-centralised ownership and independent local programming (as cited in Moran, 1992, p. 102). That is not to say that networking was non-existent. As Armstrong (as cited in Moran, 1992) infers, networking did exist but not to the extent that it occurs today.

These earlier networks relied on co-operative agreements between station owners. For example, in 1938 the Macquarie Radio Network was one of the first networks to be established, formed by member stations (Ames, 2005, p. 183). In economic terms, this made sense as stations were able to pool resources and access the top rated radio programmes, such as the *Lux Radio Theatre*, and reap the benefits of the associated advertising dollars (Moran, 1992, p. 30). This type of network arrangement also allowed radio stations to extend their audience by reaching areas outside the major cities (Ames, 2005). However, as Ames adds, this method of networking in such a highly concentrated market is difficult. As owners increase the number of radio licences they hold, they are less dependent on being a member of a network to access programmes. In terms of general programmes such as talkback, they have the ability to create their own such programmes and feed that show to their own network without paying any associated fees to stations outside their network. This situation is highlighted by the way the SRN has constituted itself differently to the Macquarie network; SRN is self-reliant and has no real need to co-operate with stations outside of its control for the procurement of programme material. For example, the John Laws show is now run by the SRN and as such hubs supply this programming to all its AM stations in the network. The SRN also has the ability to on-sell this show to stations outside of the network.

In its submission to the *Regional Radio Inquiry* in 2001, the then Federation of Australian Radio Broadcasters stated that networking proceeded from the introduction of the *BSA 1992* and that "the reasons for networking come down to four basic issues - community needs, competitive issues, economic necessity and programming variety" (FARB, 2001, p. 3). There is also the assertion that the role of networking is one that keeps regional audiences updated and informed in line with the objectives of the BSA:

> It is through the use of networked programs from metropolitan and regional areas, in combination with locally-produced content that commercial radio has succeeded in achieving the BSA's objective of diversity, responsiveness to audience needs, the provision of high quality and innovative programming and the appropriate coverage of matters of local significance in regional areas. (FARB, 2001, p. 4)

FARB claims that the BSA, a regulatory tool, assists radio stations in responding to the needs of the audience. Therefore, with a number of regional stations controlled by the one owner, it can be expected that networking of these stations will continue to occur, as is the case for the SRN, especially since competition for advertising increases from new and traditional media such as the Internet, television, newspapers and other radio stations. It has been argued that without deregulation—which allows licensees to establish networks without penalty and it seems with the express permission of regulators—regional areas may have been left without a service (ABA, 2003, p. 37).

Methodology

The research took place at four SRN stations: two AM and two FM; two of which (one AM and one FM) are the network hubs (feeder stations). The stations at the hub not only feed material to regional SRN stations, they also broadcast local programmes to their respective licence area. The two regional AM and FM stations broadcast an average of 18 hours of networked material per day. As mentioned, a large proportion of the material broadcast on SRN regional stations is sourced from a feeder station (hub). Due to the amount of networking, interviews and observations took place during the breakfast programme at each station. This is the only locally hosted programme at the majority of SRN stations and it was felt that this would provide a valid comparison of the stations. I interviewed and observed programme-makers (journalists and announcers) and managers (including programme directors). The journalists were observed between 4:30 a.m. and 11 a.m., and announcers between 4:30 a.m. and 9 a.m., and interviews were conducted immediately afterwards. The only exception was that the afternoon (FM) and drive (AM) announcers were observed and interviewed [1] during their respective programmes, as these were not only networked to other regional stations but were also broadcast to their local area. Three journalists and seven announcers participated in this aspect of the case study research.

Networking, programming and localism: city versus country

Fry (1998) and Starkey (2004) argue that towns and cities in regional areas have their own identities and a person's acquisition of this identity occurs most often from living within the community. As such, access to, and use of, this local knowledge affects the style of broadcasting, programme content and its appeal to that audience.

Prior to deregulation there was a clear objective that the provision of a broadcast licence (radio) was "intended to provide a local or regional service" (DoC, 1984, p. 25), the idea being that a local owner of the radio licence would be better situated to meet the needs of their local community.

> The dedication of commercial broadcasters to the concept of localism arises from their recognition that effective broadcasting depends on programming that is relevant to the needs and interests of the local community and which benefits their listeners. (DoC, 1984, p. 132)

In its submission to the *Review of Localism Policy in Australian Broadcasting* held by the Department of Communications (Australia) in 1984, FARB held the view that being local was integral to regional radio (DoC, 1984, p. 133). It stated:

> Licensees are very aware of its significance and, as a result, station managers and staff are constantly participating in a variety of community

---

[1] Participants involved in this research project have been allocated a pseudonym in order to protect their identity.

activities which relate to the station's place in the minds of the people it directly serves, but which have only an indirect bearing on station programming. (DoC, 1984, p. 133)

However, in its submission to *Local Voices: An Inquiry into Regional Radio* in 2001, FARB argued strongly that localism:

... doesn't have to be someone in the studio in the town to which the programme is being broadcast. It is about what comes out of the speakers from the consumer's perspective – it is material of relevance and appeal to the local audience. (FARB, 2001, p. 3)

While there are a number of regional stations that maintain a high level of local programming, radio networks are taking advantage of economies of scale offered by deregulation that allow high scale networking of content. As indicated by the current controllers list from ACMA (2009), interests based in capital cities own the major regional networks. As such, it can be argued that managers and some programme-makers at network hubs located in mostly larger regional or metropolitan centres also maintain a similar concept of localism as that stated by FARB in 2001.

The opinion of owners and the radio industry body CRA since deregulation on localism differs from most of those who work in regional radio. While many see local radio as a business rather than a community asset, interviews with regional SRN programme-makers indicate some disagreement with the current position on localism and, as such, networking. In the case of the SRN, regional programme-makers have a different perspective on local radio compared to programme-makers and managers at the hubs. As this latter group operate in competitive markets, they consider the regional networked stations inconsequential to the overall success of the hub in its own market since they appear to have made no direct financial contribution to that hub, despite the financial contribution regional stations may make to the network as a whole. This situation was reflected in the attitude of managers and announcers towards preparing programme content for regional stations. As one manager at a hub stated:

In a competitive market we probably have more say on our programming here because of ratings, but in the country [regional areas] where there's no ratings or no competition they can afford to have what they get, because it's still good programming and it's still a niche market. (Manager 2, Interview, 13 October, 2009)

Some announcers at the hub also indicated that in terms of preparing for their programmes, the network feed was not important. One announcer declared:

> Well, because I'm doing production for most of the day and I do a four-hour airshift, there's really no time to put prep [preparation of programme content] into the regional feed. (Announcer 3, Interview, 15 September, 2009)

Another announcer stated:

> I don't put any thought in it to be honest. I just treat it as though it's an audience, be it all over the place. (Announcer1, Interview, 16 November, 2009)

It is understandable then, that regional stations express a degree of frustration at the way they are treated. One announcer at a regional station was annoyed with the lack of co-ordination of the music played locally and across the network. They explained that, on occasions, a song that was played at the end of their programme was played again at the start of the network feed. From a programming perspective, and to minimise the perception that a station is receiving a network feed, this lack of co-ordination and communication means that announcers are not able to forward promote programme content.

> If you weren't aware that it was networked you'd think that we were idiots. (Announcer 6, Interview, 14 January, 2010)

This lack of care during the network feed was also expressed by management:

> If anything does go wrong up there [the hub], we're looking like fools from here. From this end the people think everything's coming from here. There's been jocks [announcers] in the past that have been on the network that aren't brilliant with their panel and they make mistakes... technology is poor. The technology needs massive overhauls. (Manager 3, Interview, 14 January, 2010)

It was also noted in the course of this research that there is no programme director[2] (PD) at the majority of the SRN stations. While the role of a PD is varied, their main role is to "develop and to direct the station's on-air sound as the company's product manager" (Ahern, 2011, p. 279).

Management at one regional station overcame the lack of a PD by consulting with "the announcers and any salesperson that might come up with an idea" (Manager 3, Interview, 14 January, 2010). However, this situation really only applied to story selection and, to some extent, music, but not the actual format of the station. The manager added, "The guys [announcers] have been told to utilise other forms of media to get quirky stories", indicating the move away from locally derived content (Ibid.). While some local managers are over-seeing their station's programming, PDs at the hubs explained that in terms of programming, they concentrated mainly on the hubs station's programmes first and then considered the programme content being delivered to the network stations. While these PDs mentioned that they provided guidance, especially to some of the FM stations, as to what music they should be playing, this was not reflected in some of the interviews with regional programme-makers. As one manager at the hub said, in terms of the regional FM stations, "The main sort of thing we get involved with is the music" (Manager 1, Interview, 9 September, 2009). However, as it was explained to this researcher, the music for one particular regional station was the same as another regional station. This suggests that while management at the hub was "involved with the music", it was more to ensure that the format was more generic than tailored to the licence area of the station concerned. Since there was no PD, the announcers chose music they felt was more suitable for the station's audience than what they were directed to play. Added to this, the placement of station IDs, commercial breaks and the like, were left up to the announcer:

> … your station promo's [promoting competitions or segments on the station] and whatever else you have becomes a format. So you play a song, you might have a station ID in-between. It's really up to the announcer. [There is] no programme director so it's up to you [the announcer] and what you think. (Announcer 4, Interview, 14 January, 2010)

---

[2] Most radio stations have changed the title of programme director to content manager or content director to reflect radio's presence online and on other platforms.

This announcer also explained that while it was their decision as to when these elements should be played, they were programmed for the network as a whole, rather than for the individual station to tie in with the break in the network feed. On the surface, this dominance over regional stations by city PDs and tweaking (the flexibility to adjust the format by the regional station) at a local level allows the regional stations to maintain some degree of localness in amongst all of the networked material. However, without a PD, there is a risk that the music and stories selected by announcers for broadcast will not relate to the licence area and are chosen on the basis that the announcer likes the story. While network owners may see no need for a PD, programme-makers did express a need for someone with programming experience to be in that role at their station. One announcer stated they would "feel more confident if I had a mentor here or somebody to give me more direction" (Announcer 6, Interview, 14 January, 2010).

Furthermore, an experienced PD in the station would also provide a clear direction for the station overall. Announcer 4 (Interview, 14 January, 2010) questioned the ability of programme-makers to determine what to broadcast if audience surveys are not conducted in the licence area. The suggestion is that without proper research or a PD to overlook what is being broadcast, it can be difficult for either the hubs or local programme-makers to determine how material on that station reflects the local culture. Without proper direction, programming tends to become a "one format fits all" approach or is treated in an ad-hoc manner. With ad-hoc programming, the announcers make decisions on what they will do as the situation calls for it. As one programme-maker said:

> I can add [music and other elements] as the programme's going if I want. There are emails that get sent around the network about what's playing well at other stations. So sometimes I'll take that into consideration. But as far as the sound of the station, there's no complete direction in where that comes from. (Announcer 6, Interview, 14 January, 2010)

While there are PDs at some SRN stations, it seems they do not mentor or share their knowledge with other regional stations in the network. Therefore this generic style of programming is seemingly the norm for networks as they can avoid employing a PD in each licence area and greatly minimise costs. This situation was noted during an observation period where an announcer pre-recorded their network feed prior to going to air at the hub. Access to generic programming like this allows the ad-hoc delivery to happen, as it was noted that most announcers were on-air at both the hub and the network feed at the same time. With this ad-hoc or generic programming in place, how do stations maintain some degree of local content? While the programme-makers at the regional and hub stations are very aware of the importance of local content, they all expressed a sense that gathering local stories was becoming increasingly difficult. A regional announcer asserted that:

> I try and get as much local content as possible. I mean it's hard in some of these towns. There's really sometimes not much going on. (Announcer 6, Interview, 14 January, 2010)

In support of this apparent predicament, a journalist at a regional station said that there are days when nothing happens and they rely on generic news releases:

> Today's one of those days [where there is a lack of local news]. So with that sense you've got to be creative. We get obviously all the NSW government press releases and a lot of the time that's generic as well. That's covering all the state and you can sort of make a story about that locally because it affects local people. You've just got to write it a different way. So when you really get stuck you can do that. (Journalist 3, Interview, 14 January 2010)

This difficulty in collating local stories for discussion on air would add some weight to FARB/CRA's claims that being local is not about hosting a programme in the town from where the station is broadcasting. But I would argue that "being local" *is* about where the programme is hosted and that it is the degree of community involvement of the staff and the station as a whole that makes it local; something that is difficult in a networking situation.

Some of the literature considered for this research stated that as newsrooms were consolidated due to networking, there were fewer journalists, resulting in a loss of local voices and diversity (Collingwood, 2008; Wilson, 2010; Prindle, 2003). While the importance of news cannot be understated in radio, especially with the immediacy the medium offers, in regional radio journalism should not be the sole focus of the discourse on localism and diversity. The announcer, advertising and other announcements are also programming elements that need to be included in the discussion. Consider also that stations promote announcers as the "face" of the radio station: the interface between the audience and the station itself. For example, the AM SRN stations have a talk-based format. In the breakfast programme, once commercial breaks are accounted for, news occupies six minutes of the hour compared with 36-40 minutes for the announcer. Therefore, it is the lack of a local announcer rather than local journalists that could just as readily threaten the flow of localised and diverse information in regional radio. Add to this the high degree of involvement of announcers in the community, and the importance of the announcer becomes apparent:

> An expectation in the older days was that a lot of people [announcers] stayed in the job a lot longer and therefore they became a part of the community. They were involved in local activities..... Your job wasn't just to come in, do the programme and be unseen. You might call bingo at the club.... you were very involved in the community... Those things can't happen under a network. (Announcer 4, Interview,13 January 2010)

Furthermore, Prindle (2003, p. 298) argues that an audience prefers a station that broadcasts local programme content. However, as mentioned, networking has made this difficult. In the majority of the regional markets that the SRN operates, both the AM and the FM licences are held by the one proprietor who lives outside the licence area and is committed to networking programme content to these stations. Therefore, it becomes difficult for the station and its announcers to make a lasting and intimate local bond with the licence area.

Conclusion

It appears that what is important to the audience is of secondary importance to the industry, which gives priority to commercial interests. Furthermore, while regulators may prescribe local content, what is actually broadcast is often left to the programme-maker. That is, while meeting the three-hour minimum of locally hosted programming, the content in that period would most likely fail any local content test, as observation of the programme-makers showed that the majority of this decision-making is either generic or nationally based, with very few local issues raised. The exception to this is news, the live reading of local weather, advertising, and community service announcements, which suggests that networks like the SRN continue to operate most of their stations at the bare minimum of the legislated requirements.

Through observations and interviews with programme-makers and managers, this research shows several issues. Firstly, it is perceived by the hubs that as the regional stations did not obviously contribute to their revenue, the on-air programmes at the hub take priority over the programme broadcast to the network. As the hub stations exist in competitive markets, their priority is to programme the hub for the market they broadcast to in order to attract an audience. This audience is then sold to advertisers to fund the hubs operations. Also, there seemed to be little compassion for any problems experienced by the regional stations when they were taking the network feed. Secondly, there was no consistency in the overall programming of the regional stations. While the stations were given a degree of direction from the hubs, they did not seem to take advantage of the opportunity their local window to broadcast a greater number of local stories would give them to create a more local feel, rather than the high amount of national stories they did broadcast. Thirdly, the operation of a network is a complex process with a diverse range of needs and wants that have to be satisfied. The hubs are located in highly competitive markets and are feeding regional stations programme content geared to a city or national audience. The only time that there is any local input during a network feed is when a person from a regional licence area contributes to a talkback programme.

This chapter discussed just a small part of the research conducted to date. Further analysis of the data will determine if networking is having a lasting effect on regional radio and its identity in the Australian media landscape. However, one can reasonably claim that the question of diversity, localism and the value of networking have created a divide between programme makers in the regional areas and those in the cities. The evidence confirms that many of the major decisions are not made at a local level but at the headquarters of the network in Sydney or the hubs. It can be claimed that this debate about localism also extends to the radio industry's representative body, CRA, and the regulator, ACMA, that appear to impose city-centric views at a regional level. While localism may never become a legislative requirement, regulators are willing to impose local content rules; albeit that they cater primarily to the needs of proprietors rather than regional licence areas.

References

Ahern, S. (2011). *Making Radio: A Practical Guide to Working in Radio in the Digital Age* (3rd ed.). Sydney, Allen & Unwin.

Australian Broadcasting Authority (ABA). (2003). *Understanding Community Attitudes to Radio Content*. Sydney: Australian Broadcasting Authority.

Australian Communication and Media Authority (ACMA). (2009). *Current Controllers: Controllers of Commercial Radio and Commercial Television Broadcasting Licenses and Newspapers Associated with Licence Areas*. Sydney: Australian Communications and Media Authority. Retrieved October 7, 2009, from http://www.acma.gov.au/webwr/_assets/main/lib100450/2009_10_06%20-%20cc.pdf

Ames, K. (2005). Hello Australia? The Impact of Networking on "Local" Commercial Regional Radio Programming. In S. Healy, B. Berryman & D. Goodman (Eds.), *Radio in the World: Papers from the 2005 Melbourne Radio Conference* (pp. 183-193). Melbourne: RMIT Publishing.

Braman, S. (2007). The ideal v. the real in media localism: Regulatory implications. *Communication Law and Policy, 12*(3), 231-278.

Collingwood, P. (1999). Commercial radio 1999: New networks, new technologies. *Media International Australia, 91*(May), 11-22.

—. (2008). *The Re-Structuring of Australian Radio, 1975-2000: Public Sphere Infrastructure in Two Capital Cities*. Paper presented at the ANZCA 08 Conference 'Power and Place'. Retrieved May 26, 2009, from http://anzca08.massey.ac.nz

Department of Communications (DoC). (1984). *Localism in Australian Broadcasting: A Review of the Policy, August 1984*. Canberra, Australia: Department of Communications.

Fairchild, C. (1999). Deterritorializing radio: Deregulation and the continuing triumph of the corporatist perspective in the USA. *Media Culture & Society, 21*(4), 549-561.

Federation of Australian Radio Broadcasters (FARB). (2000). *A Submission to the House of Representatives Standing Committee on Communications, Transport and the Arts: Regional Radio Inquiry*. NSW, Australia: Federation of Australian Radio Broadcasters.

—. (2001). *Final Submission by Federation of Australian Radio Broadcasters Limited to House of Representatives Standing Committee on Communication, Transport and the Arts: Inquiry into the Adequacy of Regional Radio*. NSW, Australia: Federation of Australian Radio

Broadcasters.

Fry, K. G. (1998). A cultural geography of Lake Wobegon. *Howard Journal of Communications, 9*(4), 303-321.

Griffen-Foley, B. (2004). Midnight-to-dawn programs on Australian commercial radio. *Journal of Radio Studies, 11*(2), 239-253.

Hendy, D. (2000). *Radio in the Global Age.* Malden, MA: Blackwell Publishers.

Johnson, L. (1981). Radio and everyday life: The early years of broadcasting in Australia. *Media Culture & Society, 3*, 167-178.

Jolly, D. R. (2007, July 24). *Media ownership and deregulation in the United States and Australia: In the public interest? 1*, 2007-08. Parliament of Australia Parliamentary Library website. Retrieved October 5, 2008, from http://www.aph.gov.au/library/pubs/rp/2007-08/08rp01.pdf

Marcato, P. (2005). Shuffling the deckchairs: Australian FM commercial radio, 1986-1992. In S. Healy, B. Berryman & D. Goodman (Eds.), *Radio in the World: Papers from the 2005 Melbourne Radio Conference* (pp. 172-182). Melbourne: RMIT Publishing.

McQuail, D. (2005). *Mass Communication Theory* (5th ed.). London: Sage Publications.

Moran, A. (1992). *Stay Tuned: The Australian Broadcasting Reader.* North Sydney, NSW: Allen & Unwin.

National Museum of Australia (NMA). (2008). *Paul Keating.* National Museum of Australia website. Retrieved October 5, 2008, from http://www.nma.gov.au/education/school_resources/websites_and_inte ractives/primeministers/paul_keating/

Neyland, D. (2007). *Organizational Ethnography.* London: Sage Publications.

Potts, J. (1989). *Radio in Australia.* Kensington: New South Wales University Press.

Prindle, G. M. (2003). No competition: How radio consolidation has diminished diversity and sacrificed localism. *Fordham Intellectual Property, Media Entertainment Law Journal, 14*, 279-321.

Shingler, M., & Wieringa, C. (1998). *ON AIR: Methods and Meanings of Radio.* London: Arnold.

Starkey, G. (2004). Estimating audiences: Sampling in television and radio audience research. *Cultural Trends, 13*(1), 3-25.

Wilson, J. (2010). The quiet north: The decline of localism and Broadband's 'Imaginary Future'. *Communication Politics and Culture, 43*(1), 70-90.

CHAPTER EIGHT

UNITED KINGDOM MUSIC
RADIO PROGRAMMING:
GOOD RADIO RECORDS
AND THE IMAGINED AUDIENCE

J. MARK PERCIVAL

A small fraction of new record releases make it onto the playlists of music radio stations. Record labels devote a great deal of time, effort and money to the production of music that will maximise their chances of achieving a place on those playlists. The questions I want to address in this chapter are: in what ways are radio programming decisions constrained by notions of the imagined audience? What is the relationship between a "good record" and a "good radio record"? How long is a "good radio record" still "good"?

This chapter draws on interviews conducted in the mid-2000s with music radio programmers in the UK, working for the most part in mainstream daytime FM radio in the commercial sector, and at the BBC. I also spoke with some independent and major label promotions personnel (or pluggers).[1] I argue that decisions on music programming emerge from a complex interaction of professional ideology, market research, and subjective judgements on the value of music as both art and commerce. All of this is framed for programmers by their sense of who is already listening and who they may wish to attract to their station, that is, the

---

[1] The interviews cited here were carried out by the author between 14 December 2004 and 30 November 2005, in Glasgow and in London. Given the potentially sensitive nature of the questions and the frankness of many of the answers given (particularly by respondents who were, and are employed in commercial radio or the record industry) the anonymity of those sources has been maintained, as promised to participants at the time of interview.

imagined audience.

The imagined audience

Notions of audience are central to both the BBC and to the commercial radio sector in the UK. This was not always the case. In 1978 Frith argued that "The BBC is contemptuously certain that Radio 1 satisfies its listeners, but it can only be so certain because its argument is circular [...] – the BBC moulds as well as responds to public taste" (Frith, 1978, p.91, as cited by Rothenbuhler, 1987, p.79). Rothenbuhler goes on to argue that in US commercial radio in the late 1980s, "except as a mental image held by the programmers, the audience enters little into programming decision making" (Rothenbuhler, 1987, p.89).[2] For UK commercial radio however, the market researched audience has been, and remains the primary constraint on music radio programming (Negus, 1993). The UK's publicly funded public service broadcaster, the BBC, imagines its audiences as the "public" to which it has a formal service responsibility under its Royal Charter (BBC, 2011). Although there are remnants of the attitude that Frith identified in 1970s BBC music radio programming, one of my arguments here, following Hendy (2000a; 2000b) is that this "audience-as-public" is central to decision making in the BBC's music radio programming. In this chapter I will address some of the ways in which audiences are imagined by music radio programmers, and how that imagined audience works to shape music programming strategies in the UK.

The perceived economic value and cultural influence of music radio is closely associated with audience ratings, both in terms of total listening figures and in the socio-economic status of listeners. RAJAR[3] ratings provide both of those sets of data to the subscribing broadcasting organisations (the BBC and the commercial radio industry in the UK) and it is this information which shapes the imagined audience. All the music programmers interviewed in this research place audience at the centre of their programming strategy. The picture they build of listeners matters in different ways and for different reasons but even those who claim to

---

[2] Revised versions of this work appear in Rothenbuhler and McCourt (1992). Another useful analysis of US music radio programming strategies is in Ahlkvist and Faulkner (2002).

[3] Radio Joint Audience Research (RAJAR) is the UK radio ratings organisation, funded jointly by the BBC and the commercial radio sector. See the RAJAR website for more information, http://www.rajar.co.uk

favour instinct or gut feeling over music research believe that they have a clear understanding of their audience's music tastes.

The expansion and consolidation of the radio industry in the UK and elsewhere has been accompanied by a move towards more conservative music radio programming. This process, described by Berland (1990) as techno-rationalisation, has been accelerated in the UK by a regulatory environment and historical context which has discouraged the development of genre-defined format radio, and therefore encourages the pursuit of a mainstream heterogeneous radio audience as a strategy for maximising ratings. In most parts of the UK then, most daytime music radio stations are competing for very similar audiences. The consequences of this are twofold: first, the same kinds of records tend to be programmed at the same times of day by competing stations; second, and more importantly, very fine distinctions between audiences become crucial, so the value of audience research increases.

There were ambiguous responses amongst my music programmer respondents to the value of music research. In some cases, programmers were defensive about suggestions that music research does not present an accurate or useful picture of audience taste. The head of music at a major London based commercial FM network said:

> We do call-out [telephone survey] research and we're now doing a lot of stuff online. We [also] put 100-150 people in a room and get them to listen to 500 clips of songs and measure [the audience reaction]. I've heard so many arguments about how music research is bollocks - well, if it was bollocks, why would it be that all across the UK, in different radio stations in different groups, we all find that the same records test [positively]?

The face-to-face music research referred to here is normally known as auditorium testing and it is commonly carried out by market research companies on behalf of radio stations or networks. Typically a radio station indicates the demographic they wish to target, based on their own preliminary market research. The market research company then finds a sample of listeners and non-listeners to music test within that demographic. The clips of music are short, usually around 30 seconds, and are typically restricted to the chorus or hook of the song. Research subjects are normally asked how much they would like to hear that song on the radio, which is of course a subtly different question to "How much do you like this song?" There are, of course, methodological issues here, as there are with any focus group research. How for example are *non*-listeners found? What is "non-listening" anyway? What does it mean to indicate on a scale of 1 to 10 how much one would like to hear a song on the radio? These are real issues but in many ways they are a distraction from more interesting questions: what do music programmers *do* with the data from this form of market research? How does it work to shape their notion of audience? And, how does that inform their programming strategy?

This head of music's point, that research works because the same records test well all over the UK, is interesting. The uniform response argument is deployed here to support the use of music research as a fair and accurate tool for helping programmers to choose records. If it is true that the same records *do* seem to test positively across regions and formats, there are two depressing possibilities, both of which have repercussions in terms of cultures of popular music and in the ways in which radio audiences hear and understand pop. The first of these is that the research methodology may be flawed at some level and tends to produce a bias towards a certain set of results, regardless of other variables. This may be the case, but the international professional association of market researchers, the Market Research Society (MRS), spends time and money attempting to reassure their members' clients that the research data produced by their members is as accurate and reliable as possible (Market Research Society, 2011). If, however, the research methodology is indeed flawed, the results produced may not be valid, and so a flawed research methodology could tend to return data that reproduces and reinforces notions of a conservative listening audience.

The second possibility is that there is nothing fundamentally wrong with music testing methodology and that audiences across the UK really *do* want to hear the same relatively familiar unchallenging sounds in their commercial music radio stations. If this is the case then mainstream daytime FM stations are correct in their tendency towards conservatism in programming. This then becomes a question of cause and effect, which is never a straightforward argument in media and cultural studies: to what extent does the conservatism of daytime music radio programming reflect the conservatism of music taste in a mass audience, and to what extent does unchallenging music programming sustain or reinforce audience taste in music? The reality is probably a complex interaction between these two possibilities and other cultural and social factors, but the effect is the same: the imagined audience is a conservative audience.

Heads of music often refer to "gut feeling" or "instinct" in making programming decisions which in this context represents their subjective experience of internalised rules constructed over many years in radio.[4] What they "know" is that certain records or certain sounds have helped their stations achieve their primary objectives, that is, increased audience in a specified target demographic, and increased advertising revenue (commercial stations) or political and cultural capital (for the BBC). Music research can provide raw data but my results suggest that programmers understand their work largely in terms of the interpretation and application of that data, often guided by professional "instinct".

The attitude to, and use of research at the publicly funded, public service station BBC Radio 1 is notably similar to that of the large commercial station groups. A Radio 1 music policy executive observed of the station's audience research:

> We ask how often they go gigging or do they go to the movies. We do call-out research once every three weeks, where we phone someone up and play them hooks down the phone. We've also been trying out online research. For us, music research is just part of that massive pool of information that we tap into. When we decide we're going to playlist an artist, it's almost despite the research. When we first started to play Coldplay we got such poor results for the first three singles, Shiver, Yellow, and Trouble. But we have a belief in certain artists; that the audience will get there. It was months and months later that we started to see [those songs] researching well. Music research is a tool, but it doesn't dictate the playlist. It is however massively important in commercial radio.

There is some ambiguity in this representation of the function of music research at the BBC. On one hand there seems to be a lot of it, including call-out research, auditorium testing and online research. This suggests that Radio 1 does indeed value highly the results of its own audience research. On the other hand there is an attempt to distance the BBC and its "public service" nature from a purely research driven approach to programming and consequently from the station's commercial music radio rivals. Research is clearly positioned as being only part of a number of sources of information used by Radio 1 programmers as they make decisions on the content of their playlists. Research here does not "dictate" the playlist at Radio 1, but at the BBC's commercial rivals it is "massively important". While there is an implicit criticism of the research driven

---

[4] For example, Ahlkvist, 2001, p. 348.

approach of commercial FM stations, my respondent was a former commercial radio programmer and may therefore be in a good position to make a judgement on the contrast between the BBC and the commercial sector. However, as a current employee of the BBC she is likely to reproduce differential arguments that positively distinguish Radio 1's popular music provision from that of commercial radio. The Reithian message here is clear: the public, represented in this instance by the imagined audience, didn't know they wanted to hear Coldplay until at least three singles into the band's major label career. It was thus Radio 1 that nudged their audience into liking what is still one of the most commercially successful bands in the world. The implication is that commercial radio would not have been able to play a part in breaking Coldplay because when records test badly there is little or no chance that they would be added to playlists.

The BBC in the form of Radio 1 is thus positioned as a taste-leading cultural organisation. Other questions emerge from this discussion: does Radio 1's decision to support Coldplay suggest that those early single releases in 2000 were good records, good *radio* records (that had so far gone unrecognised by commercial music radio), or both? And what is a good radio record anyway?

Good records and good radio records

The notion of the good radio record underlies much of my discussion here and this notion is as difficult to pin down as the idea of a truly "authentic" artist, perhaps more so.[5] While there may be any number of frames within which various constructions of authenticity in popular music can be described (Peterson, 1997), it seems that the good radio record is significantly more elusive. My hypothesis before starting the interview process was that there would be a characteristic or set of characteristics of a record that my respondents would understand as crucial to their decisions about what they judge to be good for their radio station, or perhaps radio in general.[6] What was less clear to me was the extent to which music programmers would or would not be able to describe those characteristics. While formal musicological analysis of hit records is not

---

[5] The fraught question of what "good" music might be addressed with provocation and insight by Frith (1990). While my emphasis here is a little different, some of the same issues arise.

[6] Negus (1999 and 2002) was very useful in framing my hypothesis and first thoughts on the research discussed in this chapter.

something that is part of audience or music research for radio programmers, it is often a form of musical content analysis that many of my respondents use to frame their notion of the good radio record. More frequently though, programmers struggle to express in any meaningful way what it is they feel they "know".

In many cases the strategy which allows them to answer the question is to point to examples of artists or records that have retrospectively *become* good radio records through airplay and, at least in part, through sales. Addressing that problem of expression is central to understanding the relationship between record companies and music radio and it is the dynamic of that relationship is what feeds back into the culture and sound of popular music. A head of music for a major London based commercial FM network responded as follows to the question of what makes a good radio record:

> I don't know actually. That's a really tough question. A band that we all absolutely loved was Maroon 5. We went to see them, and we were all standing watching them going 'Jesus, there's about 4 or 5 fantastic records on here that our audience is gonna love!' And sure enough, they're our biggest testing songs and we're loathe to take them off the list because they're still working. So, I don't know. It's usually... I just think 'that'll work'. I don't think you can put your finger on it. And we don't always get it right. We sometimes play something that doesn't work or we hang back a bit longer than we should have done, on something we should have played.

This programmer, like many others, feels that he "knows" when a record will be good for his network but, interestingly, he also acknowledges that gut feeling and instinct are not always correct. This suggests that the notions of instinct require reinforcement or validation, in this case by audience research. A music programmer for two regional English contemporary hits FM stations also called on notions of intuition in making playlist decisions:

> Well, it's feel. Like the new Black Eyed Peas record. I was just like, 'Bang. You're in'. It's the feel of the record, it's the hook, the way the song has been put together. And because I'm searching for a particular sound for the station, hit records aren't always as obvious as that, sometimes you need to give things time to bed in. Obviously the instantaneous things we want to get on straight away, but it can take a while for audiences to get new music and you have to really sell it into them.

There is a frustrating vagueness about programmer responses to the question of what makes a good radio record – this programmer uses the example of a Black Eyed Peas song that he immediately playlisted. This, like other examples listed by my respondents, can help work towards establishing a canon of records that have been hits in terms of sales and in airplay. The imagined audience and its response to musicological cues is central to this northern English FM network head of music:

> There are so many different factors as to why that song is or isn't going to connect to your audience. Sometimes it's the production, sometimes it's the hook, sometimes it's the perception of that artist. Good examples are really fantastic pop records by artists that I just *know* carry negative perceptions in my audience. Girls Aloud have done some great pop records but we know that Girls Aloud are not particularly liked or respected by 38-year old mothers in council estates. Somebody a little more grown up like Natalie Imbruglia or Natasha Beddingfield is far more appealing.

There is a difference in emphasis between these music programmers on their understanding of the good radio record. The southern programmer implicitly addresses audience when he refers to the "sound for the station". The northern programmer explicitly builds his notion of the good radio record around audience research. The southern programmer's response suggests that the sound of the station helps to shape audience perceptions and tastes. The northern programmer's response suggests that the imagined audience dominates programming decisions and shapes the sound of the station. In practice I believe that these are not discrete processes and that most commercial radio stations mix strategies according to market conditions, competition and corporate objectives.

The northern programmer differentiates the good radio records that appear to work for his target demographic from the good radio records that are simply "good". His research tells him that his audience has "negative" perceptions of some artists, in this case Girls Aloud, which is in conflict with his argument that the group make great pop records. The point here is that the musical characteristics of a record are only part of what makes that record good for radio. It is irrelevant that Girls Aloud songs are full of great pop hooks and that the production sound of those records has been successful with critics, record buyers and many FM stations. The research-generated notion of "artist perception" by audiences easily overrides those musical qualities for the northern network's target demographic - choosing new records for their playlist is often about finding artists with whom an imagined audience can identify. Research on how emotional attachment to artists functions is largely framed in terms of fandom (Lewis, 1992) but it would be interesting to see research that attempted to unravel the issues of attachment and identification as mediated by music radio.

It might be reasonable to expect BBC notions of a good radio record to be a little different to that of the commercial music radio industry, given my discussion earlier in this chapter and the BBC's often repeated positioning of audience-as-public. A BBC Radio music executive combined musicology, instinct, notions of professional experience and the BBC as cultural leader and public service broadcaster:

> It's something that's a bit different, something that's got a hook, something that's exciting. It might not always be my personal cup of tea, but when I first heard Antony and the Johnsons, I thought 'what the hell is this?' But y'know, when certain people you respect, like Zane Lowe, are going crazy about an artist … In hindsight now, y'know they won the Mercury, and it was very good that we playlisted them. But on first listen I thought 'Woah, I don't get this at all'. There are certain tracks that are just so obvious, like the first Arctic Monkeys track [*I Bet You Look Good On the Dancefloor*, released October 2005]. It just had such an instant sort of hook, I knew that it was going to really work. So when I listen to music, I listen with an open mind, but I do have an ear for what is really playable on Radio 1 daytime. For us it's not as simple as what makes a good radio record because our job at Radio 1 is to challenge the audience, to take risks with music. But if it's an important piece of music or an important artist, or a scene that's developing, we have to play these pieces of music that if we were a commercial radio station we just wouldn't go near.

It's not clear here how an artist or a piece of music might become understood as "important" but it appears to be at least in part how a particular artist or record is receiving attention at other music media outlets. An "important" record will only become so if it receives sufficient critical acclaim early in its life, or if later it becomes a huge hit, perhaps even a future classic. It is however politically and culturally useful for the BBC to be able to say, "We were the first to play this (now *important*) record", so Radio 1 must then attempt to stay close to sites of innovation and be aware of other media activity around a new sound or scene. It is clear though that the remit to find new music and to challenge the audience competes with other corporate objectives to build audience figures. This is not a new problem for the BBC (Hendy, 2000b) but the two examples that my Radio 1 respondent gives are instructive. Mercury Music Prize winners in 2005, Antony and the Johnsons, do not sound much like the rest of Radio 1's daytime output and my respondent's "What the hell?" reaction may be completely understandable. In the summer of 2005, playlisting Antony and the Johnsons was a brave decision, but this was an example of a critically lauded artist with less commercial cross-over potential for radio than his fellow New York night club scene band, The Scissor Sisters. The Arctic Monkeys, on the other hand, followed in the wake of Franz Ferdinand's chart success to arrive on daytime music radio with intelligent, British, guitar-driven pop songs, and were consequently much less risky for the BBC's national youth pop network.

This BBC music policy executive's experience in UK commercial radio contextualises her understanding of her role at Radio 1. It is about making decisions about which sounds are just challenging enough to fulfil the station's perceived public role without causing undue damage to ratings. How then did her notion of a good Radio 1 radio record differ from a good radio record for commercial radio?

> Interesting question. I think commercial radio probably does too much audience research, so they get a certain sound, a certain type of record. And it just ends up being that all the commercial radio stations are playing the same small handful of records. But what we're finding is that people are getting a bit bored of hearing the same records on all the radio stations. Maybe the traditional idea of what makes a good radio record isn't actually something that *is* a good radio record. But there are some records that are so obvious. Like when you hear Beyonce's 'Crazy In Love' you just know that's a great record and everybody's going to be able to play it. It's the hook, the production, the melody. If people want to listen to the radio it's usually there on in the background, they're just going about their business, so it shouldn't be something that's like this loud noise coming out of the little box in the corner. That in a traditional sense is a good radio record. Yeah, good production, simple lyrics, easy to sing along to.

This Radio 1 programmer is correct about commercial radio's use of audience research as the dominant influence on programming decisions. She refines the earlier respondent's discussion of production values and hooks with an understanding of the listening context for music radio in general – audiences are normally doing something other than simply listening to the radio. She also suggests that for the commercial sector in particular, the soundtrack supplied by music radio should not be too loud or distracting. Music radio is represented here as usually there "in the background", and so a good radio record in this context is something that doesn't unduly disrupt the "background-ness" of commercial FM music radio.

Good records and good music

If the negotiation of the notion of the good radio record has been slippery, how then do issues of taste and aesthetic preference impinge on those decisions? In what ways does it matter how programmers and pluggers make value judgements about the good-ness of music? Most of my respondents found it particularly difficult to address this question. The head of music of a northern English hits network was an excellent example of the problem of conceptualising and expressing professional practice:

> What is good music? Good music is… er, that's such a subjective thing really, isn't it? Something I do find frustrating about the industry that we're in is that, y'know, people will be very negative towards manufactured pop acts. However, quite often, it's those manufactured pop acts that are paying the wages of the record label that can afford to sign the next Radiohead.

This programmer, like most music fans, understands good music as being a subjective judgement largely unrelated to general critical discourse and critical consensus. Unlike most music fans, however, he is a former major label plugger and is now programming a network of FM radio stations. His position on good music is framed by critiquing "serious" music fan attitudes to pop acts. He understands the political economy of the record industry, which needs to generate income from current artists in order to sign and develop new artists (Negus, 1992). This makes it possible to see "manufactured" pop acts as *good* music, alongside critically acclaimed "serious" artists like Radiohead.

A senior BBC Radio 2 music executive was reflective about the difference between good music and good radio records:

> Good music is different. You can get a track that has a great hook but you know it'll only be around for six or eight weeks. That's not good music, that's a good radio record. You can have other tracks that you think are beautiful but that won't stand heavy rotation. We could play a good record once or twice and that's it, because I know the audience might like it now, but I also know that they'll get fed up with it. You just *know*. I can't qualify it.

The length of time a record remains viable on a playlist is, for Radio 2, one of the differences between good music and a good radio record. This is an oblique restatement of the traditional divide between pop (of its time, transient) and rock (timeless, classic). Hook-heavy mainstream pop records work well, even on a station like Radio 2, but this programmer implies that more "serious" music wouldn't work in the same way. There is some conflict here – it is not that the imagined Radio 2 audience does not want to hear more "beautiful" or complex pieces of music but, for this programmer at least, his audience does not seem to want to listen *often* to a record with those characteristics. The process through which those decisions are actually made is, however, shrouded within ideas of instinct and gut feeling. He attempted to clarify his point by listing artists that represent his (and by extension, Radio 2's) notion of what "good" music is:

> When I put the playlist together I don't think, oh this is young or this is old, I think this is good music. So whether it's KT Tunstall, or Madonna or Stevie Wonder, or David Gray or Goldfrapp or Bob Marley or Franz or Katie Melua. They're all on our playlist because they all make great music. The younger generation came along and they wanted something different to what their peers had been listening to. Good music and melody and songs is what they have discovered, which basically what Radio 2 is all about.

This list of long established and more recent artists underscores the sense that "good" music is "timeless", but also that there are many records which can be both "good" music and good radio records. There is also a not unreasonable assumption that Radio 2's audience want to hear "good music and melody and songs". The correlation between this programmer's music policy and the growing ratings of his station makes it possible to ignore the more complex questions of how his own notion of "good" might differ qualitatively from that of his audience or from that of other professionals in radio or the record industry. The head of music of a modern rock FM station is more direct:

> Well, I'm a music fan to be honest. It's very easy to spot what a good record is. A good record is a good record. That's my personal taste, and [programming music] is not about my taste.

The circularity of this argument is far from unusual – a good record is obviously a good record because it's good. This programmer's point though, is more that personal preference in music shouldn't impinge on his professional activity as a music radio programmer, despite the fact that he describes himself as a music fan. A music programmer for two regional English contemporary hits FM stations responded to the question of the difference between good music and a good radio record with a more elaborate version of the same argument:

It's really funny – I've got an ear for both [good music and good radio records]. My first ear is for radio. There are things that are good for radio that I personally don't like but I know that they'll work [on radio], and there are records I love that really work well on radio. There are things that I think are great pieces of music that I love to listen to at home that wouldn't particularly work on the radio.

These three categories of records are defined not by what they sound like or what values they might represent. They are rather formed by the way he processes the music he hears every day as music programmer. These ring true to my own experience, not as a mainstream FM station programmer but as a specialist music DJ with BBC Radio Scotland – all new music is passed through the filter of "Will it work on the show/playlist?" The first category (good for radio but disliked by DJ or programmer) is the one least likely to trouble the specialist DJ, yet on many occasions I played records that I considered to be interesting, important or innovative but for which I had little personal affection. For the mainstream commercial music radio programmer it is likely that the second category (good music which is also good for radio) will vary in size and content depending on the format of the station being programmed. The strength of programmers' understanding of their own professionalism works to prevent the movement of records from the third category (good music that wouldn't work on radio) to the second. From the other side of the radio/record industry relationship, an independent record plugger's perception of the value of music was more apparently straightforward:

There are two types of music in my book, good and bad. Stuff you like and stuff you don't like. Stuff I'd listen to on the radio I wouldn't necessarily put on at home [and vice versa]. If you do have a good radio record, the chances are that it's a good song. But that's not always the case - some of the biggest radio records of the year are records I don't like. Daniel Powter's 'Bad Day' for example was a number 1 airplay record; I can see why other people like it. Does that make it a good record? To other people yes, to me, no. Radio stations like it, people must like it. Whether it's good music or not is very subjective.

Despite his career in record promotion, this plugger's notion of a good radio record is anchored in his experience as a radio listener, a somewhat different perspective to that of a music radio programmer. His record industry experience is more evident in his suggestion that a good radio record is probably also a good song, even if he does not personally like that record. In this situation he is prepared to accept the proposition that if other people, in this case programmers and audiences, believe a song to be good then it probably is, despite his personal antipathy to that record.

For how long are good radio records still good?

What about records which are not new releases? The so-called "recurrent" record is the backbone of British commercial hit music radio, and plays a major part of programming policy at Radio 1 and Radio 2. Definitions of "recurrent" are fluid, as will become clear from my respondents, but in my view the most interesting aspect of the recurrent record is how rarely it is discussed in most research on music radio. This seems strange for two reasons. Firstly, the recent-hit status of recurrent records allows stations to maintain a sense of continuity for an audience – the recurrent record is familiar and its popularity has already been established. Moreover, the presence of that record on air over a longer period of time allows radio stations to establish a stronger association with an artist or sound. Those stations then continue to be the place where one might expect to hear, say, Green Day, regardless of the time elapsed since the album's release date. Secondly, the recurrent does something significant, if less obvious for the record industry – it maintains audience awareness of an artist whose album may have been available for a year or more. In economic terms this is likely to promote "long tail" sales that are relatively low on a week-to-week basis, but are cumulatively important (Gibson, 2007). Many commercial hits stations have a predictably conservative approach to refreshing playlists. The head of music of a northern English FM network said:

> How long a record stays on our playlist is an interesting question for us, because for us it's a lot longer than it is for other formats. It is possible, within our format, that something can stay on the A-list for 6 months. Sometime pluggers' views on this are positive, sometimes they're negative. It's positive because that record is constantly selling albums. But secondly, it can be very negative, because we're not playing the follow-up singles; we're staying with the first one.

Much of the programming on this particular network would be considered by other music radio stations to be recurrent-heavy and my respondent here reflects on the consequences of that for his relationship with pluggers. His assessment of pluggers' attitudes to the slow turnover of playlist may need some clarification – most pluggers are likely to find this programming approach frustrating for many of the reasons that he identifies. Long term album sales are good for the record industry, and much commercial radio programming contributes to that objective. However, the record industry has a problem when it becomes more obvious that music radio has its own, quite different, set of commercial objectives. This programmer sees a record industry that releases singles too quickly for radio because of the long-established promotional strategies of single, followed by album, followed by tour. His priority then is extending the on-air life of a single because his music testing research tells him that this is likely to increase the chances of attracting and holding his target audience demographic. He argues that this is also good for the record industry, which benefits from those longer term record sales. A major label plugger concurs:

> [Recurrent records are] very important. We don't sell records for one week or two weeks, we sell them for 365 days of the year. We know we're onto a good thing with a new artist particularly if they stay on the radio after their release ... if [radio stations] keep playing them then it means they mean business and they trust the artist and trust the record.

This plugger recognises that the record industry is not simply about short term sales and chart placings, but needs to sell music throughout the financial year. The continued presence on air of a record maintains audience awareness of an artist, which is an important part of breaking that act as a long-term album seller. Whether a recurrent record necessarily indicates radio station trust in an artist as well as the individual song is probably more difficult to assess than is suggested here, but it is reasonable to imply the connection. At BBC Radio 2 there is a rejection of formal music testing designed to establish whether a record has outstayed its playlist welcome. A senior Radio 2 music executive said:

> All other stations test for the 'burn' factor and we don't do that. When I listen to Radio 2 I *know* that if I hear a particular record much more, I'll get tired of it. So we take it off before we get tired of it. There are tracks I might rest for six months, or I might rest them for a year, but I go through everything every so often and I decide whether I should start playing a record again, or on a lower or higher rotation.

Here he contrasts his gut feeling approach to decisions about when a record becomes a recurrent against the research led approach of the commercial stations. There is an instructive correlation here between apparently qualitative decision making and the quantitative validation of that gut-feeling approach. RAJAR figures show steady growth during this programmer's tenure at Radio 2, so he may feel justified in his use of instinct in combination with audience research. He is, however, far from alone in his struggle to deconstruct his own professional instincts. There are here, and elsewhere amongst my respondents, strong echoes of Schlesinger's (1987) work on BBC news journalists, their understanding of news values, and their problems in deconstructing their own decision making process.

BBC Radio 1 has a clear focus on current and recent records, and decisions on which releases become recurrent seem to be based at least in part on audience reaction to those releases. The relatively small number of older records on Radio 1's playlist emphasises the station's public service commitment to new music - fewer old records played means more space for new releases. A Radio 1 music policy executive suggested that only a small minority of current records are ever likely to become recurrent:

> Some records when they come off the published playlists, it's like 'who cares?' Nobody's bothered about hearing [for example] the last McFly record [again], whereas there're some records like James Blunt's 'You're Beautiful' that will hang around forever. It is about the artist. But if you're got a great record you need to play it regardless. Like The Caesars' 'Jerk It Out' – we still play that. It was just a great pop record which worked well, but really, The Caesars ... nobody cares [about the band]. It was just a record that had a [TV commercial] with it. Sometimes it's just about the record. And with dance acts, that's definite – it's not about the artist. We'll be playing Bob Sinclair for ages - we get so many requests for the whistling song. But The Gorillaz can do no wrong by our audience, or Coldplay. Sometimes there's a sound that they love, like a certain artist or a producer.

For this Radio 1 programmer there is an obvious difference between artists that are likely to make records that become recurrent and those that do not, though her perception of when an audience "love" a song is probably informed by SMS, email and telephone feedback to individual shows on the network. At first glance her position appears to once again mirror long-standing music and media industry notions which represent rock as authentic (and timeless) and pop as inauthentic (and ephemeral). Throughout the history of commercially recorded music the social construction of authenticity has been central to establishing the cultural value of artists and sounds (Peterson, 1997), but since the late 1970s constructions of authenticity have become increasingly complex and multifaceted (Frith, 1981; 1996) and here my respondent illustrates that dynamic and flexible use of authenticity. Records that tend to become recurrent are the "good" records and, here at least, also "authentic" records. "Authentic" values may then be associated with either a one-off record, which is often the case in dance genres, or with "serious" rock artists developing a career over a longer period of time.

There is a far from a clear line of argument here though – the "goodness" of a record, at least in this programmer's list of examples, is not directly related to the sounds on the record. McFly (a "manufactured" guitar-based pop-rock band whose first single was released in March 2004) are considered to be so "pop" that their records are unlikely to become recurrents. A record by Bob Sinclair, a proponent of commercial dance music since 1996, is considered to be similar to one by Swedish alternative rockers The Caesars, in that they both became recurrent, but my Radio 1 respondent understands the audience as being interested only in the record rather than the artist. Those records are understood as being one-off hits, whilst the recurrent artists like Coldplay and Gorillaz will continue to release hit records that will become future recurrents.

This model of how hit records become recurrent records is relatively straightforward at the opposing ends of the pop/rock continuum. "Manufactured" pop (or rock) is less likely to become recurrent, while "authentic" rock (or pop) is more likely to become recurrent. Between those extremes the rules are significantly less clear – a dance music record from a genre associated by a number of my respondents with ephemerality (Bob Sinclair) is on equal footing with a one-off hit from an alternative rock band (The Caesars). What the Radio 1 approach has in common with that of the commercial stations is that those decisions, despite their apparent dissimilarities in ideological underpinnings, are framed by the perceived needs of the imagined audience.

Conclusions

The imagined audience, as public or market or hybrid of both, is at the centre of music radio programming decision making. The resources spent by the BBC and the commercial sector on researching radio audiences suggest that radio is audience led in both sectors. It is an understanding of broadcasters' perception of the imagined audience that illuminates the processes of music programming which is designed to deliver specific consumers to advertisers or to maximise a predetermined favourable balance between ratings and the performance of public service principles. Music policy is informed by the dynamic balance between research, instinct and ratings for all my radio programming informants. Where that balance rests is dependent on whether a station or show is regional or national, BBC or commercial, generalist or specialist.

Good radio records are sometimes also perceived as good music, but it depends on *what* (rather than *who*) the radio station is for. Music radio programmers tend to find it difficult to express what it is that they do professionally every day, but some patterns have emerged which suggest that there is a matrix of values which inform that decision making process. For commercial FM music programmers, it is in the sound of the record, the status of the artist, and in how these fit with station notions of audience. The good radio record may or may not also be good music - for my respondents there is a suggestion that a good radio record may at times not be good music at all, but only if one of the principal criteria for defining good music is its longevity, or perhaps more accurately, a record's ability to become part of a canon of classics. For BBC music radio programmers there is an additional layer of complexity in playlist decision making – the visible demonstration of the BBC's commitment to its audience as a *public* rather than as a *market*, even if this particular imagined audience still at least in part, a market. Therefore BBC Radio 1 and Radio 2 have a strong vested interest in choosing records which are not only appropriate for the station sound and image, but which are also potentially *important* records.

References

Ahlkvist, J. (2001). Programming philosophies and the rationalization of music radio. *Media, Culture and Society, 23*(3), 339-358.

Ahlkvist, J. and Faulkner, R. (2002). Will this record work for us? Managing music formats in commercial radio. *Qualitative Sociology 25*(2), 189-215.

BBC (2011). *BBC annual report and accounts 2009/10.* Retrieved May 1, 2011, from http://www.bbc.co.uk/annualreport/download/index.shtml

Berland, J. (1990). Radio, space and industrial time: The case of music formats. In T. Bennett et al. (Eds.) (1993), *Rock and popular music: Politics, policies, institutions* (pp. 105-118). London: Routledge.

Frith, S. (1981). *Sound effects: Youth, leisure and the politics of rock'n'roll*. New York: Pantheon.

—. (1990). What is good music? *Canadian University Music Review* *10*(2), 92-102.

—. (1996). *Performing rites: On the value of popular music*. Oxford: Oxford University Press.

Gibson, O. (2007, 18 July). Arctic Monkeys set hot pace in Mercury prize. *The Guardian*. Retrieved August 13, 2008, from http://www.guardian.co.uk/uk/2007/jul/18/musicnews.music.

Hendy, D. (2000a). *Radio in the global age*. Cambridge: Polity Press.

—. (2000b). Pop music radio in the public service: BBC Radio 1 and new music in the 1990s. *Media, Culture and Society 22*(6), 743-761.

Lewis, L. (Ed.). (1992). *The adoring audience*. London: Routledge.

Market Research Society (2011). *MRS Annual review 2009-2010*. Retrieved March 22, 2011, from http://www.mrs.org.uk/mrs/annualreview.htm

Negus, K. (1992). *Producing pop: Culture and conflict in the popular music industry*. London: E. Arnold.

—. (1993). Plugging and programming: Pop radio and record promotion in Britain and the United States. *Popular Music 12*(1), 57-68.

—. (1999). *Music genres and corporate cultures*. London: Routledge.

—. (2002). The work of cultural intermediaries and the enduring distance between production and consumption. *Cultural Studies 16*(4), 501-515.

Peterson, R. A. (1997). *Creating country music: Fabricating authenticity (2nd Edition)*. Chicago: University of Chicago Press.

Rothenbuhler, E. (1987). Commercial radio and popular music: Processes of selection and factors of influence. In J. Lull (Ed.), *Popular music and communication* (78-84). Newbury Park, NJ: Sage.

Rothenbuhler, E. and McCourt, T. (1992). Commercial radio and popular music. In J. Lull (Ed.), *Popular music and communication (2nd Edition)* (101-106). London: Sage.

Schlesinger, P. (1987). *Putting 'reality' together*. London: Routledge.

CHAPTER NINE

JUST BE YOURSELF?
TALK RADIO PERFORMANCE
AND AUTHENTIC ON-AIR SELVES

HELEN WOLFENDEN

There is a standard piece of advice that talk radio presenters almost always get when they start working on-air: "Just be yourself". It sounds easy, especially when you hear people every day who are good at it. But more often than not, as soon as you are sitting in the studio by yourself, trying to talk into the microphone, the words that come out and the way they come out sound nothing like you expect – or the way the advice suggests. There are exceptions to this; people who can step up to the microphone and sound as if they have been there forever. But for many people it is a struggle – and with good reason. The advice belies the complexity of the task at hand.

This chapter is based on research with practitioners about how they do their work. Within studies of radio, *practice* is generally unexplored. Hesmondhalgh (quoted in Beck, 2003: i) calls it "the vitally important but shamefully neglected topic of cultural work". Enquiry tends to focus on broadcasting outputs. However, much can be gained from engaging practitioners in developing our understanding of the medium.

Which self? Which personality?

The modern style of radio presentation has been described as "personality radio" (Geller, 1996; Guilfoyle, 2002). This refers to broadcasters who build a relationship with their audience, based on embedding their authentic self – their personality – into their on-air presentation. These presenters are highly desirable to radio managers because, the theory goes, they attract and keep audiences. The reality is likely to be much more complicated than that. Personality has always been a nebulous term and the "self" is also a tricky concept to unpack. But given that a radio presenter's livelihood will depend on it, it is worth delving into the complexities.

Some corners of academia have noted the dilemma. Tolson calls it:

> ... the 'professional ideology' of media presentation... But what do these people talk about when asked to describe the key attributes of the job? Overwhelmingly and routinely, these are reduced to the imperative of 'being yourself'. (2001, p. 446)

The essence of the problem is contained within the expression, "Be yourself". It assumes that human beings have a single identity, a single personality. "Be yourself", in the singular, does not leave any space for more sophisticated understandings of self, which have long gone beyond the unitary (Blumer, 1969; Cooley, 1922; Mead, 1934; Sullivan, 1953). So it would be reasonable for a presenter to ask: "Which self?" Added to the unitary self is the implied requirement for the "authentic" self. Montgomery illustrates the complexion of authenticity within a broadcasting context:

> Because broadcast talk by its nature takes place in the mediated public sphere, it is frequently - to a greater or lesser extent - staged for performance: and the performed character of the talk displays itself in various ways - for instance, in the pre-allocation of turns, in the reactions of a studio audience, or in a perceived sense of scriptedness. 'Authentic talk' in the public sphere might, by contrast, be seen as the reverse of this. It is a condition to which some kinds of broadcast talk aspire, in which traces of performance are effaced or supressed. (2001, pp. 397-8)

But while researchers like Montgomery and Tolson can help clarify the phenomenon, the view is that of an outsider. There has been a surprising lack of engagement with practitioners about how they experience and respond to these pressures. How does a radio presenter answer the question of how to "be yourself" on air? As part of the research on which this chapter is based, I conducted depth interviews with 14 radio presenters, eight producers, two leading radio trainers and close family or friends of the presenters. The presenters and producers all worked for Australian Broadcasting Corporation (ABC) Local Radio stations or Radio National, presenting live daily (weekday) programmes.

My research process was informed by my own experience as a talk radio presenter. I began my broadcasting career as a "rookie" with the ABC regional Local Radio network, and as well as working on my own presentation skills, was in time responsible for recruiting and then training and managing several beginning on-air broadcasters, many of whom had no specific training in broadcasting. The apparent simplicity, but practical elusiveness of a normal, ordinary, engaging and authentic presence on air remained a challenge. Standard devices or training formulae seemed to have limited connection with practitioners' actual process, and limited effectiveness in improving it. Within this set of puzzles, an investigation into how radio broadcasters actually achieve the state of being a natural, funny, attractive, interesting, intelligent conversation partner, while sitting in a room by themselves, was long overdue.

Authenticity and performance

For the first decade of the 21st century, the ABC's approach to presentation was encapsulated by the highly naturalistic imperative of "personality radio". There had been a move away from the more remote, "objective" and authoritative presentation tradition inherited from the BBC. Now presenters were being encouraged to tell their own stories, to bring their lives and experiences into the programme content, and to allow listeners to build a connection with a "real" person – to have a "conversation" with the listener. Why the shift? The authentic self, as embodied by the conversational presentation style, is attractive to radio producers and station managers, "presumably because its verbal forms project in the public sphere in a cluster of values widely held to be desirable: egalitarianism, informality, intimacy, greater possibilities for participation, and so on" (Montgomery, 2001, p. 398).

Another reason is likely to be about creating a perception of presenters as "ordinary" – a mechanism that would also bolster authenticity. Tolson draws on Sacks' (1984) notion of "doing being ordinary", with the understanding that "Being an ordinary person is not something which is pre-given" and "...'being ordinary' is accomplished in the ways people tell stories about their experiences, in typically mundane ways" (2001, p. 449).

For some of the radio presenter interviewees then, my question of whether the on-air work is a performance represented a significant challenge to the dominant discourse. Performance is not considered to be "ordinary" or "authentic". "Performance" is a loaded word, redolent of staginess or calculation.

In his seminal sociological study, *The Presentation of Self in Everyday Life* (1959), Erving Goffman dedicates his first chapter to "performance". Goffman's discussion is centred on ordinary and everyday contexts, and for most people a radio studio does not fit into that category – unless you are a radio presenter. But Goffman does take a particular interest in radio at various points in his work. And while a radio studio might be considered to be a foreign environment for most, listening to the radio, and the listening context for radio (in the car, kitchen or via a personal audio player) very much fits within Goffman's everyday, familiar setting for the evolution of a "self".

The radio studio is a very strange environment. As a presenter, your job is to sit in an often padded room, in front of a microphone and a complicated technological console, and speak to—as Paddy Scannell describes it—the "unknown, invisible absent listeners". This is a challenge which Scannell describes as the "fundamental communicative dilemma for broadcasters" (2000, p. 10). But even though the radio "product" is produced in this strange environment, for the listener who hears it in his or her everyday listening context, it must feel warm and familiar. If we understand the self as socially constructed, as Goffman and Symbolic Interactionists do (Goffman, 1959; Mead, 1934), that makes the task for presenters of working out an on-air self—or how to be yourself on radio—very tricky indeed.

Goffman defined performance as "all the activity of an individual which occurs during a period marked by his continuous presence before a particular set of observers and which has some influence on the observers" (1959, p. 22). There are, of course, other definitions. In this situation, performance can be taken to mean "inhabiting a character" or "being someone else". It may even be understood as a corollary of the pejorative description of someone being histrionic; of "putting on a performance". With this in mind, to question whether a presenter is performing when he or she is on-air may seem to contradict any effort to be "authentic".

Do practitioners think that it is a performance?

In the research, presenters' responses to questions around this issue proved complex and sometimes contradictory. To an initial question about

whether presenters consider themselves to be the same on-air as off-air,
responses included:

> "I think I'm almost exactly the same"; "I don't think very different at all";
> "The same person I am to my friends"; "In some ways yes and in other
> ways absolutely not"; "You always need to stop and think a little bit about
> what you say"; "I try to be similar"; "Much closer this year"; "In a lot of
> ways yes"; "Relatively close"; "It has to be you"; "Yes, but I think that
> there are different people, not that you not become, but that you sort of
> are"; "The very thing you need to be on the radio is yourself".

The comparison of the on-air to the off-air self is a challenge to the
"just be yourself" rhetoric. It also explores a particular kind of authenticity
– the level of congruence between how presenters recognise themselves in
their "radio presenter" social context, and the way they see themselves in
other social contexts. For many, there is an on-going struggle to
approximate one to the other. For others, there is a resolution of that
struggle in the awareness that multiple selves are employed, both on-air
and off, depending on the way the social context shifts.

Social theory has long held this to be a complex matter. "Erving
Goffman has shown the constructed nature of identity, the self as a
presentation or performance designed to be appropriate to the
circumstances and settings in which it is produced in the presence of
others" (Brand & Scannell, 1991, p. 201). What is important to recognise
here, however, is that off-air, the appropriate self deployed to match social
circumstances is much more automatically drawn out or elicited by the
social context. By comparison, the on-air context, at least initially, has
much or all of that information missing. What a presenter is required to do
is appropriate his or her understandings from off-air social contexts to
apply on-air. This takes time, as we hear when one presenter describes
herself as "much closer this year".

But it is not a straight translation from off-air to on. There are
modifications needed to the off-air self in order for it to meet the needs of
radio presentation and avoid transgressions. The most obvious of these is
swearing – but there are other functions required of the presenting role that
must also be accommodated. These include time calls, personal and station
identifications, as well as simultaneously-performed technical functions,
such as operating the radio desk and "timing out" so the content fits
precisely within the allocated programme duration. Added to that is the
difference between the social context in which the talk is produced (the

radio studio) and the variety of social contexts in which it is received (perhaps the car, kitchen or anywhere else you take your audio player). This changes again once you factor in variables such as programme content and purpose (for example current affairs programmes as compared to companionable general interest shows), time of day, duration of the programme, geographic location and the frequency with which that programme is presented. And then within each individual programme, the social context potentially changes from moment to moment. The act of taking talkback calls condenses social contexts such that a presenter may be required, within the space of seconds, to move from talking to an elderly lady to a truck driver.

So make no mistake, this is a difficult task. It is not a simple process to know which parts of your off-air selves you want to plagiarise for this particular social interaction and how they have to be modified and tweaked to meet the special requirements of the on-air context.

The presenters I interviewed provided a spectrum of responses to the question of whether on-air presentation was a performance. There were emphatic no's, equally unequivocal yes's, and those who sat somewhere in between or were still wrestling to find a definitive answer. We will start off with those in the "no" category.

Madonna King is the presenter of *Mornings* on 612 ABC Brisbane. This was our exchange on the topic:

HW: Do you think of the on-air work as a performance?

MK: No I don't and I know that's probably the wrong answer. I know other presenters think that is the case. If it is, it's not what I want it to be.

HW: What is it if it's not a performance?

MK: It's doing a job. It's doing a job that I love and that job is to tell people who are listening to me what is going on and how will it affect them. I don't know anything in that that smells of performance and to do it in a way that's yourself but to me that's not performance either. ... The first year I thought I was in an eisteddfod and it was (dramatic voice) "Good morning, my name is Madonna King!" (laughing). I saw myself up on a stage just with no audience that I could see, probably none at home either! Whereas now I know it's more a conversation. (M. King, interview, February 1, 2008)

In terms of Goffman's definition, King is clearly concerned about her "continuous presence before a particular set of observers" and she worries herself that she is not answering the question correctly. What King has done is modify and adapt the presentational self as she has become more familiar with the environment. But for King, performance has too many negative connotations to be available to describe what it is she is trying to do. Instead, "performance" describes the incorrectly applied self we might call "eisteddfod Madonna".

Lindy Burns presents *Drive* on 702 ABC Melbourne. Burns is also adamant that this is not a performance:

> LB: No. Because I did drama at uni and so I know the difference. I know the difference of... how you feel. I've been a musician. [...] I don't see what I do as a performance because I don't think it's actually all about me. It's about me giving the opportunity for people to hear from guests that we bring in who have an opinion about something that's going on at the time and for them to talk and for them to express. So it's more... I see myself more as a conduit - more so than a person who is sort of the star. I never see myself as the star.

> HW: But isn't it you that people have the relationship with and that they turn on...

> LB: Yeah that's the thing I don't quite get - apparently that is the case. And that's lovely but I find it a bit bizarre to think that they kind of love me for that because they don't actually know me...

> HW: But don't they, if they...

> LB: If I'm putting myself on the air? When I was talking about concealment before, if I'm really grumpy, you would tend not to go on air and go 'I'm SOOO grumpy today I can't begin to tell you'. So yes there would be things like the horrible aspects of Lindy I would tend to try and conceal. So... they don't get the full picture. They don't get to see what my husband gets to see, for example. (L. Burns, interview, January 24, 2008)

Burns had responded to the question of whether she was the same on-air as off by saying she thought she was "almost exactly the same". Yet Burns also says that she finds it strange that her audience loves her, because "they don't actually know" her.

I pursued this further because as the conversation developed (and this happened in many of the interviews) the inherent contradiction became apparent. Burns is clearly aware of the tension. However, not unpacking the contradictions too carefully is almost a defence mechanism. Burns, the girl from humble working-class and regional Newcastle origins who now holds a highly prized on-air position in urban and chic Melbourne, sees performance as "being the centre of attention". By avoiding confronting that issue she keeps herself unaffected—and perhaps even "authentic" — or at least "Newcastle Lindy" authentic.

Several of the presenters demonstrate this tension. Participants would often start out in one place, and end up in another – exploring the territory as their position emerged through the discussion. It is hardly surprising, because much of the training literature avoids confronting many of the uncomfortable inconsistencies that come from digging beyond the platitudes (Geller, 1996, 2000; Guilfoyle, 2002; Mills, 2004; Trewin, 2004; Simons, 2007).

Scripting, authenticity and "fresh talk"

There are other ways in which both practice and understanding of this issue fall short of the complexities of the actual experience of on-air talk. Scripting is one of them. Some in the ABC advocate working without scripts, but it has remained contested territory. For those who use them, several questions emerge. How can you claim to be authentic if you are not being spontaneous? What if a producer has written a script for you? Montgomery contrasts Goffman's notion of "fresh talk" with "naturally occurring talk" and quotes Goffman's complication of the term "speaker".

> One meaning, perhaps the dominant, is that of *animator*, that is, the sounding box from which utterances come. A second is *author*, the agent who puts together, composes, or scripts the lines that are uttered. A third is that of *principal*, the party whose position, stand and belief the words attest. (Goffman in Montgomery, 2001, pp. 399-400)

Goffman defines "fresh talk" as speaking that "is formulated by the animator from moment to moment, or at least from clause to clause. This conveys the impression that the formulation is responsive to the current situation in which the words are delivered" (1981, p. 171). But if you thought you could solve the problem by ad-libbing, Goffman also notes that "Fresh talk is something of an illusion of itself, never being as fresh as it seems" (p. 172). Clearly in an endeavour to produce fresh talk a presenter may be any combination of animator, author or principal – but not necessarily all three at once.

Jon Faine produces fresh talk by working largely unscripted. Faine presents *Mornings* and *The Conversation Hour* on 702 ABC Melbourne. Faine is something of a stalwart in the ABC and several participants reference him in their interviews and consider him a role model.

HW: Do you think of the on-air work as a performance?

JF: Oh there's no doubt it is… Geoff Rush was on *The Conversation Hour* one day and at the end of it we sort of had a bit of a chat and you know I was star struck and terribly excited and he said "No no no. What I do," - this is Geoffrey Rush speaking, he said – "someone writes a play and I learn it. I rehearse it for several weeks and then I perform it for maybe an hour and a half in front of three or four hundred people, night after night after night, for a season. And I think that's hard." He said, "But what you do, no one writes anything for you, you don't have a rehearsal, you perform for three and a half hours live, in front of hundreds of thousands of people, and then you do a completely different show the next night, the next day." (J. Faine, interview, January 23, 2008)

Faine is able to contrast his own performance with that of an actor – even better, he is able to have the actor, who is one of Australia's finest, do the job for him. This exchange between Faine and Rush can be considered in the context of two people who are at the top of their respective crafts, contrasting the different elements of performance in each of their practices. Faine demonstrates the authenticity of his performance by highlighting the freshness of the content as well as the talk and the unrehearsed delivery. Faine would generally consider himself to be the "animator" and the "author" of his performance, but not always the "principal". He recognises that the nature of the role means that sometimes he "has to ask the mongrel question". Faine says, "I'm performing a role. I don't mean performing a role theatrically, I mean performing a role in society. It's… the ABC's obligation and role of keeping people accountable in decision making…" He says that he is "not a belligerent person but on-air [he] can be" (ibid).

The exchange between Faine and Rush does not acknowledge the "routine" and "episodic" nature of radio presentation. Brand and Scannell employ this framework, along with Goffman's notion of socially constructed selves and performance, in their examination of the work of radio presenter Tony Blackburn (1991, p. 201). Within this framework, we can see that Faine, and every radio presenter, develops familiar or safe territory within the programme. The programme will follow a similar format each day. In Faine's case, this is something along the lines of hard news early, some regular or recurring spots and guests, and the more relaxed *Conversation Hour* as the end. The elements of unpredictability and risk are inversely proportionate to the familiarity of the space, and it is reasonable to claim that the more desirable kinds of "authenticity" become more available as a presenter becomes more relaxed and familiar with their programme.

Your "best" self? Other-directedness and performance

Richard Fidler is the presenter of *The Conversation Hour* for 612 ABC Brisbane and 702 ABC Sydney, and *Afternoons* for ABC Brisbane. Fidler recognises that "we're all like lots of different people in the one" and that the performance element is being "your best self".

> RF: The self that's kind of had its cup of coffee. The self that is in a good mood. The self that's interested and hearing properly and ready for some fun, ready to just hear what anyone has to tell you. Yeah… that's the kind of performance if you like. Prepping yourself so you're in that good frame of mind. Reaching a kind of sweet spot within yourself, where you're ready to be in that frame of mind. (Laughing) And I'm not always successful at that Helen, I freely admit. […]. Yeah. They're the worst days actually. Not really with the technical mishaps, where you just churn in a dull dull show and you haven't even been able to interest yourself in it you know. You feel this kind of low level of shame.

> HW: So how do you cope when you are having a bad day and is that something you would talk about to the audience?

> RF: Then you go into performance mode. It's much more of a performance then so you fake, you put on, you do fake at that point yeah, to some degree. […] If you were to go on-air and go "Oh look I'm feeling really kind of feeling bored and tired at the moment". What's in it for the listener there? Nothing. I mean you could admit it and it would be honest but the listener's thought then is well you know, "fuck off" or "crank it up". You know "I'm here I need something from radio right now. I need to know what's happening in town, I need to feel diverted or distracted or just entertained or informed" all those things. If you're not up to it, you know, go home. So you do need to fake it on some days, some days, not often, not

often but yeah you do you fake it sometimes. (R. Fidler, interview, February 1, 2008)

Fidler gives us another example of the requirements of the role and subsequent limitations to authenticity. It is clear that there is definitely more to this than "just being yourself". As a radio presenter you have a job to do, an obligation to the people who have bothered to switch you on. If you are in the chair, no matter what your personal dramas are at that point in time, you still have to meet your responsibilities to the audience and of course, the organisation who is paying you. It is a more complicated "set of observers" that a presenter has to serve than just the "invisible absent listener". Fidler also recognises that the fact that "it could all go horribly wrong at any given time" is part of the authenticity of the performance (ibid).

James Valentine is the presenter of *Afternoons* on 702 ABC Sydney. Valentine offers a particularly articulate description of what is happening on-air:

JV: You've always got to think about it from the point of view of the listener. Here's a person stuck in traffic with an AM radio in their dashboard. What does this sound like? And so unless you're thinking about it in that sort of perspective all the time then you're not going to be creating interesting things that come out of the dashboard and soon as you're thinking like that you're thinking as a performer thinks. That's what performers think like. (J. Valentine, interview, January 29, 2008)

According to Valentine, you effectively "perform as yourself" which is not the same as "being yourself". "Performing yourself" further complicates the notions of authenticity and naturalness inherent in the "be yourself" injunction. Valentine points out that it takes time to learn how to do it; to become familiar with this strange social context and to work out an appropriate self for it. He says "With all of these sort of things, what... increases is your base level. The more you do it the higher your base level gets." He also recognises that the presentational self must at some level, be other-directed, because it is only through reciprocity that the presenters needs are met.

JV: I've got the biggest ego in the world, you know, but I also know that that ego's not going to get served unless I'm there for the audience and unless I understand what the audience is wanting. And unless...it's all about them. If I make it all about them I get my jollies. (J. Valentine, interview, January 29, 2008)

Staying amateur... and being professional

Lucky Oceans is a two-time Grammy award winning pedal steel guitarist and presenter of ABC Radio National's world music programme *The Planet*. Oceans agree that the "base level" improves but he laments the price that is paid:

> LO: When I first became a musician I noticed that my perception of my good gigs, like they were absolutely fabulous. My bad gigs were just terrible, terrible, terrible, terrible. And a lot of that is self-perception you know. The audience will see this narrow range and you'll see a huge range in your performance [...] and the same thing goes in broadcasting. But as you go on you learn the skills to deliver a decent performance in any situation you know. But you don't have that hanging on the edge feeling. And a few years ago I played with a guy named Liam Gurner and he was like 19 and I said "That's the feeling!" You never know when it's going to run totally off the rails and because of that, when it's good, it's fantastic [...]. What was it that William Blake said about, you know, you have innocence, experience and innocence regained? So that somehow that ties in with the amateur thing is by not thinking of myself as a professional, yeah I'm going to make mistakes and maybe have highs and lows, you know good programmes and bad programmes, but I'd rather have that than a sort of a cookie cutter predictable everyday show. (L. Oceans, interview, February 18, 2008)

So we have a paradox: Valentine, whose practice of the craft over time creates a performance that becomes more professional and for him, is "not very different at all" to the person he is off the radio. And Oceans, who seeks to retain some amateurism in an effort to recall a "rawness" that is also considered to be authentic.

It is important to consider the implications of the relationship between time and performance. Talk radio presenters in the ABC are often on-air for shifts of two hours, some as long as six. Over this daily duration, five days a week, forty weeks a year, it is difficult to sustain a self that is highly alien to the selves used in other social contexts. Oceans' example of performing a gig is quite a different context to presenting a regular live radio programme. Nevertheless, his challenge to "professionalism" offers another perspective by which to understand both performance and authenticity.

At the time of the research interview, Geraldine Mellet was presenting 720 ABC Perth's afternoon programme. Mellet also agrees that this is a performance.

> GM: Yeah I do. I do think of it as a performance. I don't think that I'm inhabiting a character. And I'd really, really hope that other people don't think that because that would go directly contrary to what I try and do on air. But I think, one of the frustrations I have sometimes is with people assuming that the kind of work we do is simply just sitting there and chatting [...] Thinking about what I'm going to say, preparing questions, thinking about a structure for the interview, yeah that's a performance for me. It's not just lobbing up and sitting in my lounge room and chatting with a friend. And I hope it doesn't have the negative connotations of performance that I am therefore extraordinarily different and it's a different beast. (G. Mellett, interview, February 18, 2008)

Mellet extends the boundaries of performance to include the preparatory work involved in being on air. Consequently, she comes much closer to Goffman's broad definition of performance. Mellet qualifies her understanding of performance by eschewing the definitions that are associated with negative characteristics like artifice and pretence.

Conclusions

Most people would not hesitate to label live talk radio presentation as a performance, but when the question is taken directly to practitioners, a more complicated picture emerges. The word itself is highly compromised, and therefore not available to some to describe talk radio presentation. Tolson recognises it as a "type of public performance, but a performance which, crucially, is not perceived as 'acting'" (2001, p. 445). I push that even further and argue that from a presenter perspective, it is critical that the presentation does not sound like performance – and in some cases cannot feel like a performance either.

So what is it? It is clear that for these presenters, there is an active projection of the self for the audience: a "best" self, a self at the top of their form. As in all such presentations of the self, the projection is a function of the relationship, and what the presenter would like the relationship to be, and what it will be allowed to be by their audience. The relationship is not of a friend or confidante or family member or new-person-you-met-at-a-party, though no doubt presenters cannibalise any or all of these for the purpose at hand. The relationship is of broadcaster to audience. The audience is known through the presenters' own history in the community of listeners, through conversations with talkback callers, outside broadcasts, and the sheer imaginative cast of emotionally-intelligent minds. This audience is understood and related to in the same instant as individual and community, and in the constantly shifting play of gender, class, culture, geography, in-group and out-group nuances within a conversation which is sometimes actually two-way, but is more often a complex and reflexive interactive process in which the audience can only be imagined.

Presenting public service talk radio is not a straightforward process. It is not simply a matter of "being yourself". The requirement to be "authentic", the strange social context of a radio studio, and the discrepancy between that environment and the one in which the talk is revived, mean that radio presenters take on a significant challenge.

More generally the research interviews indicate that a range of tensions emerge about the way people operationalise an on-air self. There is also a deep instability in the way practitioners think about and talk about their practice, a significant underdevelopment of discourse about the practice and how it is accomplished. This indicates a real need for researchers in this field to be working with practitioners to elaborate and clarify what this fascinating interactive process is about.

References

Beck, A. (2003). Introduction: cultural work, cultural workplace - looking at the cultural industries. In A. Beck (Ed.), *Cultural work: understanding the cultural industries*. London: Routledge.

Blumer, H. (1969). *Symbolic interactionism: Perspective and method*. Berkeley: University of California Press.

Brand, G., & Scannell, P. (1991). Talk, identity and performance: *The Tony Blackburn show*. In P. Scannell (Ed.), *Broadcast talk* (pp. 201-226). London: Sage.

Cooley, C. H. (1922). *Human nature and the social order (Revised edition)*. New York: Charles Scribner's Sons.

Geller, V. (1996). *Creating powerful radio: A communicator's handbook*. New York: M Street Publications.

—. (2000). *The powerful radio workbook*. Nashville: M Street Publications.

Goffman, E. (1959). *The presentation of self in everyday life*. Garden City:

Doubleday Anchor Books.

—. (1981). *Forms of talk*. Oxford: Basil Blackwell.

Guilfoyle, D. P. (2002). *Certificate IV in broadcasting radio learning guide: Section seven.* Sydney: New South Wales Department of Education and Training.

Mead, G. H. (1934). *Mind, self and society*. Chicago: University of Chicago Press.

Mills, J. (2004). *The broadcast voice*. Oxford: Focal Press.

Montgomery, M. (2001). Defining "authentic talk". *Discourse Studies, 3*(4), 397-405.

Scannell, P. (2000). For-anyone-as-someone structures. *Media, Culture & Society, 22*(1), 5-24.

Simons, M. (2007). *The content makers:Understanding the media in Australia.* Camberwell, Australia: Penguin Books.

Sullivan, H. S. (1953). *The interpersonal theory of psychiatry*. New York: W.W. Norton & Co.

Tolson, A. (2001). "Being yourself": The pursuit of authentic celebrity. *Discourse Studies, 3*(4), 443-457.

Trewin, J. (2004). *Presenting on TV and radio*. Oxford: Focal Press.

CHAPTER TEN

FORESIGHT, FUDGE OR FACILITATION? THE MAKING OF UNITED KINGDOM DIGITAL RADIO POLICY 1987 - 2008

TONY STOLLER

This chapter considers the emergence of digital radio policy in the United Kingdom over the past 20 years, looking at how public policy in broadcasting actually gets to be made. Why are certain decisions taken? And why does that happen at one time rather than at another? What are the considerations which motivate governments? Are they driven more by external influences, or by their own instincts? And what happens when they seek to intervene directly in market processes, to fix the market? Or to put it as you might do when you look at digital radio in the United Kingdom, how on earth did we get here?

For the years between about 1987 and 2008 we have in the emergence of digital radio policy in the United Kingdom a fascinating case study about such policy-making. There is a large amount of primary research material to work with. That embraces not only the range of published reports but also the files and archived documents of the Independent Broadcasting Authority and the Radio Authority.[1]

There are three different categories for broadcasting policy decisions. The first, which objectively we would think the best, is that careful research and creative thought foresees what the audience and the market will need in future years, and harnesses developing technologies to the best public outcomes for those needs, while protecting and carrying

---

[1] The other key data source is the recollections of those who were the policy makers. The author was one of those, as the Chief Executive of the Radio Authority from 1995 until 2003.

forward the best aspects of the existing arrangements. We may call that the 'foresight' mode.

Next, there is the type of policy which emerges on the hoof, randomly from the chaos of day to day events and systems, which might be dubbed 'fudge'. The third category of policy driver may be called 'facilitation'; shaping policy to meet the demands of external parties other than the end users, often called 'vested interests'. The emergence of digital radio in the United Kingdom is a valuable case study to examine all of those at work, and to see which in the end has proved dominant.

Digital radio policy in the United Kingdom has gone through three distinct phases so far: getting to first base, mostly in the early to mid 1990s; trying to move digital from the wings to somewhere close to centre stage at the turn of the century; and then developing the concept of analogue switch-off from 2007 onwards. The first two periods are legitimate areas of study for the media historian. The last is rather more current affairs, but there are some data of value to illustrate the later stages of the 20-year story.

This chapter will look, by way of a prologue, at the United Kingdom's very early—and never subsequently changed—commitment to Digital Audio Broadcasting (DAB). It will then consider the formation of policy at the three critical points, each of which is linked in with United Kingdom primary legislation: the 1996 Broadcasting Act, which introduced digital radio into the United Kingdom; the 2003 Communications Act, which flagged the end of attempts to use broadcasting as part of the social liberal responsibilities of government; and lastly the run-up to the 2010 Digital Economy Act.

The context for all of this is the fast-changing United Kingdom political scene. The 20 years of detailed policy-making have taken the United Kingdom through very different administrations: the neo-Thatcherism of John Major's Conservative administration from 1992 to 1997; the neo-social democratic New Labour governments of Tony Blair and Gordon Brown from 1997 to 2010; and now the new British coalition experiment under David Cameron (2010 - ?). It has all happened within three very different regulatory philosophies for radio broadcasting: the last years of Reithian paternalism; the brief flourishing of a mixed social economy for radio; and the eventual triumph of market liberalism.

The United Kingdom adopts DAB

The emergence of digital radio as a transmission possibility came about as United Kingdom radio was getting used to a true duopoly. The BBC radio had emerged from the review named *Broadcasting in the Seventies* with a concentration upon five national networks, one of them, Radio Five Live, significantly for the digital story on AM only. The BBC's collection of local radio services, almost all in England, was very much secondary. Prestige came from the two speech networks of Radio Four and Five Live; the three music networks were engaging in – and mostly losing – a battle for share of the radio audience. An increasingly successful private radio sector in the United Kingdom had originally emerged as a strange mix of entirely local, commercially funded services carrying extensive public service obligations. After 1990 it began to shift more towards what the rest of the world understands as commercial radio and added three national stations, two of those on AM only. In terms of governance, the BBC remained a formidable monolith; the private sector was regulated from 1990 by a sector-specific regulator, the Radio Authority (Stoller, 2010).

From the very start, 'digital radio' in the United Kingdom effectively meant only the Digital Audio Broadcasting technology – DAB. Although other technologies were discussed – and, in other continents actually deployed, for the satellite services deployed in the US via Sirius and WM, and in Africa and Asia by World Space – the prevailing orthodoxy in Europe was that DAB was the only game in town for European domestic broadcasters. The first CEPT radio conference, convened in Copenhagen on 13 and 14 November 1991, set in motion the process towards international agreement for the allocation of DAB frequencies.

DAB as a whole was the product of a hunger among British broadcasters for more spectrum. BBC engineers at their Kingswood Warren headquarters had been working on digital radio and digital television systems for some time. In 1987, the Eureka 147 digital audio broadcasting project was established, with an initial consortium of a few European broadcasters, many equipment manufacturers, European car makers and transmission companies. The only United Kingdom member was the BBC. In 1994, the industry watchdog in Europe, the European Telecommunications Standards Institute (ETSI), formally accepted the DAB standards, confirmed at a world level in 1995 (Stoller, 2010).

For the BBC, their substantial part in the development of DAB, and leading role within the Eureka 147 consortium, kept their field of vision pretty narrow. The Radio Authority was hugely influenced by the BBC, as was the European Broadcasting Union (EBU), where the BBC always has great influence. The EBU's Technical Director, Phil Laven, had been a leading part of the BBC's DAB project before leaving for Brussels. In turn, the EBU's preference further confirmed the Radio Authority's Chief Executive, Peter Baldwin in his commitment to Eureka, and the BBC and the Radio Authority together promoted it to the United Kingdom government accordingly.

There is no evidence to show that any other technology received serious consideration. In Europe, the state broadcasters in France and Germany were experimenting with DAB transmissions, although the private radio sector there showed little or no interest. In the United Kingdom however, the hybrid arrangement which underpinned ILR (and subsequently Independent National Radio, INR) ensured a significant residue of public service attitudes, which included interest in technological research and development at an early stage. The United Kingdom commercial radio companies' association, AIRC, had established its own DAB committee which first met on 30 August 1991. Their concern was that existing licensees "should be given the opportunity to provide their programme services via DAB" but that "stations should be allowed to simulcast DAB programmes with their AM and FM programmes until they determine sufficient DAB penetration has been achieved to rely on it as the sole transmission medium".[2]

Confirmation from government that DAB was to be their choice came from the President of the Board of Trade, Michael Heseltine, who established a United Kingdom DAB Forum in February 1993, to work on the necessary frequency planning and to encourage field trials. The language of his Parliamentary statement is revealing, showing elements of foresight but chiefly facilitating the demands of the BBC and the Radio Authority.

Digital audio broadcasting—DAB—is recognised as a very important development in sound broadcasting, and I am anxious to ensure that its benefits and opportunities are made available in this country. I am

---

[2] Letter from Peter Jackson, Chief Engineer Capital Radio to Martin Brebner, Broadcasting Policy Section, Radiocommunciations Agency 2 September 1991 (Stoller, 2010).

therefore establishing a United Kingdom DAB forum to co-ordinate and promote plans for the introduction of DAB in this country and to liaise with those with similar interests in other countries, particularly elsewhere in Europe. Participants will include broadcasters, equipment manufacturers, retailers, service providers and other interested parties. The aim is to enhance the potential benefit for United Kingdom industry and ensure that British people will be able to enjoy the benefits of DAB, including high-quality reception in cars and on portable receivers, and additional services that the new technology will make possible. My officials will continue their efforts to secure sufficient radio spectrum for the first terrestrial DAB transmissions to begin when the first receivers become available, which is expected to be around 1995.

The BBC ran its first DAB trial in London on 6 September 1993, and the United Kingdom government's technical arm, the Radiocommunciations Agency, set about clearing the frequency needs in an international conference at Weisbaden in July 1995, under the auspices of the International Telecommunications Union (ITU). Television in the United Kingdom would vacate the part of the FM band which had been used for the old 405-line black-and-white transmissions. The United Kingdom's successful pitch was to use some of this space for DAB, allowing sufficient frequency space for seven frequencies: one for national BBC services; one for national commercial services; and five to allow the necessary intricate pattern to bring local digital services–BBC and ILR–to most of the United Kingdom.

Neither at this stage nor indeed until September 2010, was there any significant listener or consumer element in these decisions. Policy development here was aimed at facilitating of the demands of the United Kingdom radio industry, the BBC and the Radio Authority, with some support also from the EBU and the acquiescence of the private radio companies; allied to such foresight as the opportunities of this so far untested technology could offer. It was strongly influenced by key individuals – Phil Laven and Peter Baldwin among others – and given credence by the technological enthusiasms of the dominant government minister of the day, Michael Heseltine.

1996 Broadcasting Act – where United Kingdom digital radio really began

The events surrounding the 1996 Broadcasting Act comprised the first act of the DAB drama. As John Major's approached its end, after a troubled five years, it turned to legislate for digital broadcasting in the United Kingdom as part of its hoped-for legacy. On 19 September 1995 it published a White Paper, *Digital Terrestrial Broadcasting: the Government's Proposals*. The Broadcasting Bill which was to give it effect came first to the House of Lords on 14 December 1995, and was eventually enacted in the very last moments of the Major administration in July 1996.

This was essentially television legislation. When the Secretary of State, Virginia Bottomley, introduced the second reading of the Bill in the Commons in April 1996, she spoke of standing "on the verge of a new broadcasting revolution even more significant than the change from black-and-white to colour television... It can bring more income to the broadcasting industry and more jobs across the country. Digital television offers improvements in picture quality, increased potential for wide-screen broadcasts, interactive television, more subscription services and, above all, many more channels and greater choice." The government's goals for digital radio were extra channels and extra choice, better quality and greater diversity of output.

The official in the Department of National Heritage (DNH) directly responsible for piloting the 1995 Bill through the United Kingdom Parliament, Paul Bolt, recalls that "digital radio was largely a 'me too' in the 1996 Act, both from the Government's and from the industry's perspective".[3] As with later legislation, it was an alliance between a senior civil servant, Bolt, and an active junior minister, Richard Inglewood (Baron Inglewood), which was the real axis in making things happen. On digital radio policy, even at this critical moment, government was relatively agnostic.

So far as DNH was concerned, it seemed fair/right to allow radio, as well as TV, the opportunity to go digital and make provision in the legislation for a framework in which that could happen. We were far from clear what the unique selling point/killer application would turn out to be and how

---

[3] Interview with Paul Bolt, Department for Culture, Media and Sport, 25 November 2010

competitive a market would develop within it. So we tried to create a reasonably flexible framework which could be adapted to market and technological developments, while always having regard to the plurality and diversity considerations which were very salient throughout the passage of the Bill. We recognised that it would take a long time for switch-off ever to be a runner for radio, but felt that in principle that should, as with TV, be the end goal.[4]

Clearly then, the legislation which actually introduced DAB into the United Kingdom was not driven by any great enthusiasm for the new technology, nor any especial foresight about how the myriad of obstacles to its success might be overcome. At the Radio Authority, this seemed to be essentially an exercise in facilitation-let's make it possible and then see what happens.[5]

However, in one respect the government was prepared to twist the arms of the far less enthusiastic commercial radio companies. The BBC and the government were well aware of the contemporary success of the independent radio sector. It was attracting at that time a greater share of all radio listening than the BBC: 49.7 per cent as against 47.2 per cent. (RAJAR) Unless the commercial companies would sign up for DAB, it would go nowhere. And why should the commercial radio companies bother to come on board with DAB? For them, this was an uncertain technology, with no evident public demand, which would simply mean doubling their transmission costs for no additional revenue. As was to be the case throughout the creation of digital radio policy and legislation, the commercial radio industry was chiefly concerned with how far appearing willing to adopt the new transmission technology would be a useful bargaining chip as they pressed for de-regulation for their analogue services.

For a couple of years, officials in Government and those at the regulator had debated providing some incentive for the companies. On 7th December 1994 the Radio Authority's top brass gave dinner to the DNH Permanent Secretary, Hayden Phillips, at which he agreed to "listen to ideas for an incentive package (which might include extending the period of licences) to encourage take-up of the new technology."[6] Four months

---

[4] Ibid.
[5] Author's observation from his contemporary experience – see footnote 1
[6] Letter from Hayden Phillips CB to Peter Baldwin, Radio Authority 19 December 1994

later, Secretary of State Virginia Bottomley announced that Government "were to guarantee the INR stations a slot on the national commercial digital multiplex, and to give [national and] local stations who took up a place on the relevant local multiplex an automatic eight-year extension to their analogue licences" (as quoted in Stoller, pp. 281-2).

This offer to INR and ILR of automatic renewal of their analogue licences in exchange for signing up to DAB, was the perfect 'persuader'. It effectively locked in the major radio companies to apply for DAB licences. However, it also forced the entire industry to become apparent supporters of DAB, when some public commercial scepticism would have been useful. The clever political 'fix' denied the introduction of DAB the wider policy debate which it needed.

Like many fixes which seem clever at the time it was also to have unforeseen consequences. The ILR companies which enthusiastically gobbled up the DAB opportunities to ensure automatic renewal of their lucrative analogue licences found later that the way in which the legislation was framed meant that they could not lay down their digital commitment without losing those analogue licences. This was to become a huge factor ten years later, when the commercial radio industry found itself in deep trouble, and it may significantly have influenced the wish of later governments to find some way of keeping DAB going lest the analogue companies were driven to collapse.

And that was how United Kingdom got digital radio, in the form of DAB. The element of facilitation, of accommodating vested interests, was the driving factor, but – the licence-renewal fix apart – these were not commercial interests. It was the keenness of the BBC and the commercial regulator, the Radio Authority, that most encouraged government to introduce digital radio into the United Kingdom in 1996. That plus a sense that if it was happening for television – where the arguments for doing so were evident and strong – it would be a shame not to tag on radio as well.

From 1996, the BBC deployed its own multiplex to simulcast its national channels, providing better-quality transmission for Five Live, and allowing BBC World Service a domestic audience. Later, in 2002, after its own internal struggles, it launched five additional national digital-only services. The Radio Authority set about from 1997 advertising and licensing 46 national and local digital radio multiplexes across the length and breadth of the four home nations offering a total of 280 programme services, and accommodating the BBC's local radio services.

Yet DAB stubbornly refused to fly. During this period, digital radio was making progress chiefly through its presence on the digital television multiplexes and the internet. The BBC under Christopher Bland and John Birt looked coolly upon digital radio, preferring the attractions of digital television and the internet, and for a while it was only the steady stream of new commercial licences which kept DAB afloat. One key problem was the lack of DAB receivers. The first portable 'kitchen' DAB radio for under £100 was not available in the United Kingdom until 2002. An Ofcom report in 2004 noted that:

> The first DAB radios were in the hi-fi, kitchen–portable and personal portable product segments, which addressed less than one quarter of the total radio receiver market. This narrow offering, combined with the relatively small range of models available within each category and the large price premium for DAB radio receivers over analogue radio receivers, constrained take-up (Ofcom Phase 1, 2004).

More than even this, despite all the policy initiatives no single killer application had emerged for DAB. In 2002 it seemed to all of us to be something for the very long term.

The Communications Act 2003 – a new digital dispensation?

The second act was the new digital dispensation made possible by the Ofcom Act 2002 and the Communications Act 2003. The five communications regulators, including the Radio Authority were swept away and replaced with one new communications super-regulator: the Office of Communications, or Ofcom. Among several fundamental shifts in government's approach to broadcasting was to give primacy to the creation of a "dynamic market", which was seen to be "fundamental to securing choice, quality of service and value for consumers". It was also to provide "a key input to the United Kingdom's international competitiveness" (A New Future for Communications, 2000).

This was unprecedented in British broadcasting, and marks the triumph of market philosophy in the media sector. Never before had competition been the first and foremost consideration, but from that point on broadcasting policy was to be driven by principles more usually applied to telecommunications and the spectrum market. Was this foresight, or just the facilitation of the new Blairite interests? Either way, it ought to have had a significant impact on the level of policy support for DAB.

Government's thinking was substantially conditioned by the extent of technological convergence between the delivery of television and telephony, coming back together more than a century after Marconi and Edison. Its timing was driven by Telecommunications Directives coming out of the European Commission. Its policy was informed by its market faith. "Developing and sustaining a dynamic market is one of the government's key objectives for helping to sustain and develop this important industry. Competition is vital to dynamic markets." DAB was seen as radio's entry into this new converged utopia. (A New Future for Communications, 2000)

As a converged regulator of supposedly converged technologies, Ofcom was drawn naturally to digital radio. Its founding Chief Executive, Stephen Carter, wanted to take bold steps to drive digital radio forward. He felt that DAB had been around for too long without making a break-through, while the bureaucracy had "allowed people to make a meal of the process" (as quoted in Stoller, 2010, p. 289). In December 2004, Ofcom published its provisional thoughts and plans for consultation, in *Radio – Preparing for the Future.* It was confident that "digital radio offers significant benefits to both broadcasters and listeners, including greater choice, enhanced services such as on-screen programme information, ease of use and reduced audio interference."

Ofcom's policy-making was dominated by its elevation of market data above everything, including the type of individual judgments which had underpinned all previous policy work. Nowhere is this approach more evident than in the changed approach to making radio policy after 2003. Ofcom itself produced seven major radio reports in five years. It is significant that when asked to describe their organisation's approach to making digital radio policy the Ofcom executives (alone among all those approached for this study) declined to be interviewed but instead simply listed their previous publications.

The years between 2004 and 2008, when Ofcom held unchallenged responsibility for developing policy in this field, produced more statistical data about radio in the United Kingdom than had ever been seen. This was an attempt to elevate 'foresight' above all other considerations. There was a clear reaction against the intuitive policy-making which had operated in the preceding years (arguably since the launch of radio in Britain in 1922). Policy was effectively led by market economists who had no prior knowledge of radio regulation.

However, during this period Ofcom's radio responsibility, the commercial radio sector fell from grace to a remarkable extent (Stoller, 2010). What was more, over this time digital radio endured a series of shocks which data-driven regulation was ill-equipped to manage. The most dramatic confrontation began in 2004, when Ofcom announced that it was to use some newly-released VHF Band III spectrum to advertise a second national commercial DAB multiplex. The GWR Group operated Digital One on what had been originally (in 1996) been dubbed the "first and only" commercial national multiplex. Faced with the threat of this new competition, and as they saw it dilution of the digital offering, GWR went ballistic.

The legal arguments continued for almost three years. GWR asserted that it was entitled to rely upon the statements of the Radio Authority that there would not be any other competing national commercial multiplex. Ofcom argued that no such commitment could be binding when circumstances changed, a decade afterwards. Digital One's Chairman, Ralph Bernard, alleged that his company had been "seduced" into investing in digital radio, having been assured it would be the sole national multiplex operator in that format. When Ofcom announced on 21 December that it would indeed advertise a further national digital multiplex, Digital One threatened judicial review. It dropped that only late in March 2006, when it claimed that it had got "got certain assurances from Ofcom", a claim which was hastily denied by Ofcom the following day (quoted in Stoller, 2010, p. 290).

In the following years, evidence accumulated of the imminent failure of this second stage of digital radio policy in the United Kingdom. In March 2008, the chief executive of one of the two major new commercial radio conglomerates, GCap, announced that the company now saw its future on broadband and FM only. "The majority of people who are listening through DAB receivers are listening to stations that are simulcasting on FM. The majority of DAB receivers out there are FM-enabled too ... If you put that against a background of the cost structure of DAB, it cannot be an economically viable platform" (as quoted in Stoller, 2010, p. 290).

On 5 July 2007 Ofcom awarded 4 Digital the disputed second national commercial multiplex licence. 4 Digital had proposed to launch the second national multiplex in July 2008. This was to be the new flagship, the great DAB hope. But in October 2008 the company announced that it was pulling out of DAB entirely, and Ofcom's statistical-led policy lay in tatters. Foresight, at least as exercised by the discipline of market economics, had failed as a policy driver.

Digital Economy Act 2008 – towards analogue switch-off?

That might well have been the end of DAB in the United Kingdom. That is was not (or at least, has not been so far) leads on to the third act, the return of digital radio policy from the regulator to government, and the impact of new government initiatives. Although it may be thought that this moves from history into current affairs, a description at least is needed to complete the story so far even if it is too early for the historian to draw any firm conclusions.

One significant effect of the 2003 Communications Act had been to shift broadcasting (and telecommunications and spectrum) policy work away from government to the new converged regulator. Under it first Chief Executive, Stephen Carter, Ofcom had established from the very start a heavy-weight and extensive policy capacity, which effectively denuded the Department for Culture, Media and Sport. Ironically, as Lord Carter of Barnes and the newly appointed minister for communications, Stephen Carter's approach to the development of digital radio policy involved re-balancing that situation, so that government could put itself in a position to take initiatives to move forward digital radio which were beyond the regulator.

Government was not to be left behind in the data stakes, but in practice the real driver for policy development was to come from a few committed individuals. Once again, as in 1995/6, there was an important alliance between a senior civil servant in John Mottram, who was one of the key "committed individuals" in this process, and his minister. Although the BBC was understandably occupied with what it saw as the primary challenge of contestability for the licence fee revenues, its Director General, Mark Thompson showed a willingness to lead a new drive for DAB. The BBC's Director of Radio Tim Davie was "a powerful and constructive force and without whom the Digital Britain chapter on Radio would and could not have been written".[7] The commercial radio industry still primarily wanted deregulation, but was prepared to see digital radio as a bargaining chip within any broader negotiation.

The Digital Radio Working Group (DRWG) had been established in November 2007 by the Secretary of State for Culture, Media & Sport, chaired by television industry stalwart Barry Cox. Its stated purpose was to bring together senior figures from the radio industry and related stakeholders, under an independent Chair, to consider three questions: what conditions would need to be achieved before digital platforms could become the predominant means of delivering radio? What are the current barriers to the growth of digital radio? What are the possible remedies to those barriers? (DRWG Interim Report, June 2008). It generated headlines in June 2008 by its proposals to 'migrate' all analogue radio services to digital by 2020. The subsequent full report, published in December 2008, brought the possible date forward to "at least 2017" (DRWG Final Report, December 2008).

The DRWG, however, had no mandate to prescribe or force a solution. The key exposition of government policy comes instead in the White Paper, *Digital Britain*, published in June 2009, of which the radio chapter is an evident segue from the main thrust of the document. As Carter acknowledges, it was: "while we're on the subject of digital broadcasting, here are some thoughts about radio...". The echoes of 1995/6, where radio was described as an "add-on", are strong.

Nevertheless, the White Paper did not shirk the task of setting out a new policy approach to digital radio. "If radio is to compete in a Digital Britain then it must have greater flexibility to grow, innovate and engage with its audience, and in this the limits of analogue as the primary distribution platform for radio are now all too visible". In the government's view, "the biggest barrier to radio's digital future is a lack of clarity and commitment to the DAB platform...Government has a pivotal role in securing this certainty" and therefore committed itself to an aspirational analogue switch-off date of 2015 (Government White Paper, 2009).

At that stage in the government's deliberations the decision regarding digital radio could have gone either way. On the one hand, it could have concluded that in the world of 3G, the internet and the like, there was no point in investing too much policy capital in a standalone transmission

---

[7] Interview with Stephen Carter [Lord Carter of Barnes] 3 December 2010

system for radio. On the other, having a separate and dedicated platform for radio in the spectrum would help the medium to feel significant. It was that latter course that won the day. Once again, radio policy occupied no time beyond those who most cared for it. Carter recalls that "in the whole process of governmental and cabinet approval of the Digital Britain report and the draft Bill, no one (elsewhere within government) asked a single question about radio."[8]

Did this follow inexorably from the data, as the post-modern approach seemed to dictate? Even allowing for the extent to which the data would have been conditioned by the terms under which it was commissioned (for example Carter explains that Myers' localness report was set up in order to produce the context in which regulatory relaxation could be offered to the commercial companies as a *quid pro quo* for renewed commitment to digital radio) that is clearly not the case.

There seemed to be little chance of renewed European drive for DAB, with European experts arguing that "radio's future may more likely be based in a multi-platform, multi-media base" (Shaw, 2010, p.235) while other nations are more likely to deploy the updated DAB+ technology[9] which is incompatible with the United Kingdom's DAB receivers. Equally, the arrival of digital radio in to the United Kingdom had not been informed by listener inputs. The DRWG, established as the main collaborative and consultative initiative for digital radio, included three government departments or regulators, three broadcasters, four trade bodies, one management consultant, and not a single representative of listeners or indeed advertisers or their agencies (DRWG Interim 2008).

The first major consumer-based report was only published in September of last year. The report by the government-sponsored report from the Consumer Expert Group emphasised "the needs and concerns of radio listeners [which] will be absolutely central to our approach to Digital Radio Switchover". It casts doubt on switch-over by compulsion, welcoming the new Secretary of State's decision to lessen the significance of the target date, stressing that only "if, and it is a big if, the consumer is ready we will support a 2015 switchover date". The report note that setting a date, or a firm commitment to a date, would have had the effect of scaring consumers to switch; that this would not be compatible with Government policy to support a switchover when enough listeners

---

[8] Ibid.
[9] As Australia has now done

voluntarily adopt digital radio; and hoping that government's new emphasis on consumers should provide the focus to ensure consumer concerns and needs regarding digital radio are addressed, thereby reducing the barriers to voluntary take-up (CEG 2010).

On any objective reading of the data, a firm move towards analogue switch-off at an early date was not indicated. United Kingdom government policy towards digital radio from 2008 onwards arose most of all because of the keen interest and enthusiasm for radio from a few key players, notably Carter, Davie, Mottram and Myers. Government concluded that "by introducing a hard switch-off date, the radio industry could be re-assured that some regulatory relief was on its way." This was essentially facilitation rather than foresight. Then, just as throughout the whole of its history in the United Kingdom, digital radio happened as it did because some people cared a lot about it and no one else cared enough to question or challenge.

Conclusion

So, was the development of digital radio policy in Britain over those 20 or so years a result chiefly of foresight, fudge or facilitation? It seems clear that foresight based on an objective reading of independent data can mostly be ruled out. The only genuine effort to do that was during Ofcom's stewardship of policy, which was notably unsuccessful for radio. The initial selection of DAB as the digital transmission technology was the decision of a few individuals which was then facilitated by the government of the day, which had itself no clear vision for how digital radio would evolve in the United Kingdom. The first legislation was similarly enabling rather than directive. The shift to competition as the major plank of broadcasting policy in 2003 was again to facilitate a political philosophy, which did nothing for DAB. The approach adopted under the current *Digital Britain* approach is unquestionably of a piece with all this, giving DAB what is surely a final chance to make its way towards permanence.

Is it fair then to consider that the policy-makers have fudged it? There is no evidence for that. But what is demonstrated time and again is that digital radio policy in the United Kingdom had moved forward as the direct result of the commitment and work of a few individuals. As the author has noted, "*we* got DAB because *they* wanted it". At no stage has there been any significant un-stimulated consumer demand for DAB. It became and has remained the central plank of the United Kingdom's policy platform because it was the wish of the policy makers themselves. To extend from this specific case study to the general, it is more than arguable to conclude that in the United Kingdom at least—and probably elsewhere too—that is how broadcasting policy is made.

References

A new future for communications. (December 2000). DTI and DCMS, Cm 5010.

Digital Britain government white paper. (June 2009). Cmnd 7650.

Digital radio switchover: what's in it for consumers? (14 September 2010). Department for Culture, Media and Sport Consumer Expert Group.

Digital Radio Working Group interim report. (June 2008).

Digital Radio Working Group final report. (December 2008).

Digital terrestrial broadcasting: The government's proposals. (September 1995). Cmnd. 2946.

Radio in digital Britain. (March 2009). Ofcom Submission to the Government.

Radio Joint Audience Research (RAJAR). www.rajar.co.uk.

Radio - Licensing Policy for VHF Band III, Sub-band 3. (October 2005). Ofcom.

—. (December 2005). Ofcom.

Radio - Preparing for the future (Phase 1: developing a new framework). (December 2004). Ofcom.

Radio - Preparing for the future (Phase 2: Implementing the framework). (October 2005). Ofcom.

Radio: The implications of digital Britain for localness regulation. (July 2009). Ofcom.

—. (April 2010). Ofcom.

Shaw, H. (2010). The online transformation: how the internet is challenging and changing radio, in Ala-Fossi, M., Jauret, P., Lax, S., Nyre, L., O'Neill, B. & Shaw, H. (eds.), *Digital radio in Europe: technologies, industries and cultures.* (pp. 215-236). Bristol: Intellect Books.

Stoller, T. (2010). *Sounds of your life: tyhe history of independent radio in the United Kingdom.* New Barnet: John Libbey Publications.

The communications market: digital radio report. (August 2010). Ofcom.

The future of radio: Localness on analogue commercial radio and stereo and mono broadcasting on DAB. (February 2008). Ofcom.

The future of radio: The future of FM and AM services and the alignment of analogue and digital regulation. (April 2007). Ofcom.

The future of radio: The next phase. (November 2007). Ofcom.

CHAPTER ELEVEN

LOW POWER FM IN NEW ZEALAND: A SURVEY OF AN OPEN SPECTRUM COMMONS

BRENT SIMPSON

Introduction

Low Power FM (LPFM) is one of the most under-researched forms of media in New Zealand. In late 2010 I sent out a self-selecting survey via two online groups, the New Zealand LPFM User group on Yahoo! Groups and the Radio Heritage Foundation email newsletter. This chapter provides a historical overview of LPFM in New Zealand and uses the results of the survey and my own experience in LPFM in New Zealand to help sketch a broad picture of the stations and services operating in this space. Minimal regulation of LPFM spectrum in New Zealand has created a form of open spectrum commons. While commons approaches can be innovative and experimental spaces, they are also subject to the tragedy of the commons.

> "The major positive impact of deregulation on the business of radio was just that – radio became a business, not just a recreational pastime for *enthusiastic amateurs*." (Shanahan & Duignan, 2005. p. 41. Author's emphasis)

In 2003 I became an "enthusiastic amateur" by volunteering at a local radio station, *The Beach FM* on Waiheke Island[1] in New Zealand. *The Beach FM* broadcast on a full-power commercial frequency, had an established brand, and was owned by a well known former New Zealand radio personality - Barry Jenkin, a.k.a. "Dr. Rock". The station stopped broadcasting in early 2008 and the frequency (99.4 FM) came up for

---

[1] Waiheke Island is a small island with a population of 8,000 near New Zealand's most populated city, Auckland.

tender in a 2008 government spectrum auction and was sold to an Auckland investor for $380,000 - a price far out of the reach of Dr. Rock and other interested groups and individuals on Waiheke Island.

By October of that same year there were three Low Power FM stations broadcasting on Waiheke Island. Two of them were "hobbyist" stations, *Waiheke Wireless*, broadcasting an automated playlist from a popular cafe in the main township of Oneroa; and *Splash FM*, an adult contemporary automated station run by a former radio engineer. The third station, *Waiheke Radio*, is a not-for-profit community radio station and media organisation run solely by volunteers.

Ten years after the introduction of Low Power FM in New Zealand, hundreds of enthusiastic amateurs are creating thousands of hours of radio in this spectrum. The creation of an open commons regime over part of the low-power radio spectrum has created a space in which localized and independent forms of broadcasting have emerged, adding to an already interesting history of radio and spectrum in New Zealand, initially controlled by government, and then secondly, by international media conglomerates and overseas capital in an open market.

Low Power FM, ownership and the broadcasting environment

The fifth Labour Government of New Zealand was sworn into office on the 5th of December 1999, coming to power on a wave of dissatisfaction with nearly a decade of free-market economic reforms. These reforms had been driven by the National government but had their origins in a radical market-based restructuring of the economy by the fourth Labour Government a decade earlier under finance minister Roger Douglas. "Rogernomics", as Douglas' restructuring came to be known, was an abrupt and sweeping range of policy changes that included the cutting of agricultural subsidies and trade barriers, privatising public assets, and controlling inflation through monetary intervention. Many in Douglas' own party saw these changes as a betrayal of traditional Labour ideals. By the election of the fifth Labour Government ten years later a new philosophical approach was being adopted by many in the party based on principles of "Third Way" politics being championed at the time by Bill Clinton in America and Tony Blair in the United Kingdom. Broadcasting Minister Steve Maharey, a leading proponent of the paradigm, sought for the establishment of much closer partnerships between state and civil society organisations, and rebranded the "Third Way as Social Democracy" (Maharey, 2001). Citizen engagement and increased access to communications technology were essential in Maharey's vision, as he articulated it to the Foundation for Policy Initiatives in Auckland in 2001:

> The implications are clear – government will be required to promote universal access to information and communication technology to ensure there is equal access to information and thereby prevent the stratification of society into haves and have nots.

Shortly after this speech the Ministry of Economic Development's (MED) Radio Spectrum and Broadcasting Policy Group released a discussion document entitled *The Future of the FM Band* which would eventually lead to the development of Low Power FM broadcasting in New Zealand.

For over a decade communications policy treated spectrum solely as a tradable commodity in an open marketplace. It is crucial to note the impact of this program on spectrum claims by the indigenous Maori people. Zita Joyce argues that the earliest Maori claims to spectrum came about in direct response to deregulation of radio spectrum and the reframing of spectrum as a saleable object (Joyce, 2008). Maori claims to spectrum rights resonated with Maharey's broader vision of a more open and fair communications regime. On March 9th, 1999, twenty days before the Crown was preparing to auction the rights to manage the radio spectrum in the two gigahertz range, Maori claim *Wai 776* was lodged by Rangiaho Everton concerning the government's monopoly right to trade radio spectrum without prior Maori consultation. Everton's claim was unique, arguing that spectrum was a space where the local cultural transmission of knowledge provided tangible communicative and economic benefits to a local population. This local position was in direct contradistinction to the increasing commodification of spectrum by the government which had opened up the broadcasting market to international ownership. Conceiving spectrum not solely as a tradable commodity but also in terms of cultural communication in the public sphere aligned well with Maharey's conception of a "mixed economy of broadcasting" in which cultural, social and commercial values of communications technology would intersect.

The majority of radio broadcasting in New Zealand is highly developed as a financial commodity and is largely controlled by a co-operative duopoly[2] consisting of on the one hand APN News and Media (APN) in partnership with the Australian Radio Network (ARN), and on the other MediaWorks, owned by Australian private equity corporation Ironbridge Capital. APN's parent company is Clear Channel Communications in the USA, the largest radio station owner in that country with over 1,200 stations in its portfolio (Rosenberg & Mollgaard, 2010). Ironbridge Capital is an Australian private equity investor that largely invests in buyouts of medium to large businesses. In 2007 Ironbridge achieved a full takeover of MediaWorks which allowed it to delist the company from the New Zealand stock market (Drinnen & Niesche, 2007).

The duopoly has also affected much of the content of New Zealand commercial radio as well. There is little musical diversity in the radio offerings available from the bulk of high powered stations in New Zealand as both companies battle over key commercial audience niches. Brendan Reilly (2011) has recently observed that even with the adoption of a voluntary 20 percent quota for New Zealand music, the amount of New Zealand music played on the *ZM* network (broadcasting in 19 markets across the country) is still relatively small with only 9.8% of all music originating from New Zealand and over half of that coming from just five artists. Seventy-four percent of the musical content across *ZM* originated from the United States and the United Kingdom.

It was arguably the lack of options around ownership, the relatively low amounts of local content, as well as the Maori claims[3] that inspired Maharey to consider, after fifteen years of deregulation, facilitating Low Power FM broadcasting in 2001.

The General User Radio Licence

LPFM in New Zealand operates under the General User Radio Licence (GURL); a self-applied licence to broadcast radio content at very low

---

[2] The two companies have joint ownership of The Radio Bureau, a research and sales agency that provides services to commercial radio and are the major players behind the commercial industry lobby group, the *Radio Broadcasters Association*.

[3] The Waitangi Tribunal was established in 1975 by the Treaty of Waitangi Act 1975. The Tribunal is a permanent commission of inquiry charged with making recommendations on claims brought by Maori relating to actions or omissions of the Crown that breach the promises made in the Treaty of Waitangi.

wattage (currently 1 watt). The licence was established in 2002 after the Ministry of Economic Development's Radio Spectrum and Broadcasting Policy Group released a discussion document entitled *The Future of the FM Band* in 2001. This document sought input into the use of "Guard Bands" at the lower frequencies on the FM band from 88.1 to 88.7 (7 frequencies), and the higher band from 106.7 to 107.7 (11 frequencies). These bands were essentially wasted spectrum, but could be suitable for low power broadcasting that would not interfere with the commercial spectrum between them or the aviation and civil uses outside the band. The Broadcasting Policy Group considered three alternatives for licensing Low Power FM in New Zealand:

> Utilisation for "in-fill" coverage in order to expand the coverage of high and medium power FM broadcasting in some areas; for "very modest" non-commercial broadcasting. The low power could ensure a true 'local' flavour at low cost; operation of a "very modest" commercial broadcasting service on a localised basis. Again the low entry cost could be attractive to small-scale broadcasters.

A large number of respondents to the document expressed an interest in using portions of the Guard Band for both commercial and non-commercial localised low power services. The existing full power community broadcasters in New Zealand, the "Access" stations, argued at the time that these frequencies would not provide sufficient coverage to warrant New Zealand On Air (NZOA) funding as they seemed worried that such localised stations could quite possibly create content complying with Section 36C of the Broadcasting Act 1989 and thus be eligible to compete for already limited funding for community radio available from NZOA. This legislation requires that the government supply the means to ensure that a range of broadcasts are available to provide for the interests of women, youth, children, persons with disabilities and minorities in the community including ethnic minorities. Currently NZOA will only consider stations with potential audiences of over 50,000 as eligible recipients for such funding, and Access stations have continuously lobbied for this seemingly arbitrary listener quota to be retained.

Nearly all respondents considered that there was a need to manage the use of such frequencies, to ensure they were being used for their intended purpose and did not exceed power restrictions. There were mixed views on whether these frequencies should be made available free of charge, or through allocation of licences by auction (as is done with New Zealand's commercial frequencies) or in return for a fixed fee. The Radio

Broadcasters Association (RBA), the political lobby group for commercial broadcasters, argued against open access to spectrum, lobbying instead for a fee system to provide for compulsory registration of LPFM operators and an initial compliance audit along with regular reviews.

In April 2002 the GURL came into effect. Radio broadcasting in the Low Power FM band was to be made freely available, in an openly accessible spectrum commons, and with restrictions around wattage and liability. The current restrictions are small enough to list in full here:

- Frequencies only within the Guard Band: 88.1 to 88.7 and 106.7 to 107.7
- 1 watt maximum output
- Within a 25 km radius of any broadcast transmitter there must be *no more* than one low power FM transmitter broadcasting substantially the same programme (including simulcast or re-transmission) as that broadcast transmitter.
- Only transmissions that are broadcasting, as defined in the Broadcasting Act 1989, are permitted.
- Low Power FM transmitter operators must broadcast the contact details of the person responsible for the transmissions at least once every hour.
- Frequency use is on a shared basis and the Chief Executive does not accept liability under any circumstances for any loss or damage of any kind occasioned by the unavailability of frequencies or degradation to reception from other transmissions.

Frequency use on a "shared basis" established the LPFM spectrum commons. In response to the frequency interference problem the government suggested that "coordination between users can minimise the risk of interference between services", suggesting user groups or associations be established to "coordinate installations for the equitable utilisation by all users of the available frequencies" (MED 2004). The implementation of the GURL opened up a small space for highly localised broadcasting to emerge across New Zealand, creating a political and social experiment in the adoption of an open spectrum commons for radio broadcasting.

LPFM and the open commons

There is no definitive source for information or statistics on LPFM stations in New Zealand. The broadcasting of LPFM requires no formal process of registration with the government or any official regulatory agency. Currently the best two sources of information on existing stations are the *NZ Low Power FM Broadcasting Community Open Source Knowledge Base*, which lists 261 stations, and *The Radio Heritage Foundation* which estimates that there are 250 to 300 active broadcasters. These two sources are for the most part user-generated and contain defunct station listings and out-of-date information. While an accurate count is very difficult to produce, a reasonable estimate, considering the available GURL spectrum and previous usage, would be around 150 LPFM stations in operation at any one time. It is however significant that so many New Zealanders have embraced the opportunity to create local broadcasting in one of the most deregulated and commercially saturated radio environments in the world.

The *New Zealand Radiocommunications Act, 1989* ushered in an almost fully deregulated open market for the licensing of high-power spectrum rights in New Zealand, initiating an often talked about but rarely implemented experiment in the use of auctions to distribute private property rights over spectrum. Winners of spectrum auctions gained twenty year ownership rights over channel assignments, yet the winners of such assignments were also free to transfer or trade property rights as well - at the time an unprecedented step worldwide.

The auction process, for the most part, favours existing media organisations with deep pockets as well as incumbent owners. In the first auction in July 1990 for 164 AM and FM frequencies, of the five available Wellington[4] FM frequencies, the average winning bid was $478,845, with the lowest going for $120,111 and the highest for $821,001. The total for all tenders in the AM-FM auction for that year was NZ$ 4.755 million (Meuller, 1993). On Waiheke Island (part of the Auckland radio market, the largest in New Zealand) the frequency 99.4 FM, which had been originally engineered to cater for the island's residents, sold for $380,000 in October of 2008 to be used to broadcast in Mandarin to the growing Chinese populations of Eastern and Southern Auckland. The expensive fees paid for entrance into much of the New Zealand commercial spectrum market, combined with an extremely limited availability and strong

---

[4] The second most populated city in New Zealand.

requirements for establishing non-commercial community access stations in New Zealand, has meant that public access to radio broadcasting is extremely difficult.

By contrast, the introduction of the General User Radio Licence in 2002 has facilitated access to radio spectrum free of charge and enabled the development of a variety of small to medium sized stations and services to emerge across the country. These stations consist of an eclectic range of broadcasters, from hobbyists to fully fledged community radio operations; Christian broadcasters to university based radio training stations. On one end of this spectrum is the lone enthusiastic amateur: generally providing an automated and specialist playlist to a small niche audience. *Chomsky FM*, for example, a now defunct LPFM broadcaster that emanated from the Grey Lynn neighbourhood of Auckland, broadcast speeches and talks by the linguist, philosopher, and political activist Noam Chomsky 24 hours a day. While at the other end of the spectrum there are full community based stations like *Waiheke Radio*. This community based non-profit consists of approximately 26 unpaid volunteers. These volunteers contribute over 120 hours of time to the station per week, nearly the equivalent of three full time positions. *Static FM*, organised and administered by Auckland University of Technology's Communications degree has up to 40 students and staff providing programming year round. Of the eight stations that identified in my survey[5] as "community stations" (none of which were on the official NZOA list of community radio stations) the average number of volunteers was 18.

In spirit the GURL was introduced to enable access to spectrum for those who were unlikely to be able to afford it under an auction process dominated by wealthy incumbent broadcasters. This has not stopped some independent locally based operators from attempting to create localised commercial broadcasting models. Chris Diack, a long-time broadcaster from Southland continues to run *Classic Gold Radio* often remotely from a mobile studio in an iconic New Zealand caravan equipped with a LPFM transmitter and a generator. Diack's mobile LPFM station recently assisted the Brighton community in Christchurch which was badly affected by serious earthquakes in 2010 and 2011 by broadcasting local civil defence and other community based messages on the radio (Cassandra, 2011). Diack has long promoted his type of LPFM radio as a "community

---

[5] In late 2010 I sent out a questionnaire about LPFM broadcasters via two online groups, the *LPFM User group* on Yahoo! Groups and the *Radio Heritage Foundation* email newsletter and received 18 responses to 19 questions.

commercial model". Two thirds of the stations in the survey identified as being run as not-for-profit; although in some cases this seemed to imply that while they were trying to make profits, they could not. Four stations were incorporated societies, and four stations were administered by charitable trusts.

But, the freedom and openness of LPFM in New Zealand has also allowed for owners of full-power commercial frequencies and stations to occupy LPFM frequencies. These have been for such purposes as to recruit and develop local volunteer DJ talent for eventual paid work on the commercial full-power station; or as in-fill coverage, neither of which is illegal under the licence. This type of behaviour has received a fair bit of criticism among LPFM groups in New Zealand. A regular topic for discussion is debating amendments to the GURL that would prevent existing holders of full-power spectrum from utilising space in the LPFM band.

In the more densely populated centres across New Zealand there can be considerable jostling and frequency interference between competing LPFM broadcasters. An example of the types of problems that can arise from the open commons approach to LPFM occurred in New Plymouth in 2008. In October 2008 the *Taranaki Daily News* ran an article entitled, "Tune in, turn on, get miffed", a story about a local LPFM frequency interference situation that was occurring in New Plymouth. The complainant had been broadcasting for over two and a half years, and was listened to by "taxi drivers, rest home residents and older people who appreciated the old style music and absence of advertising" (Scott, 2008). This station started having its signal overpowered by a nearby station, *One Christian Radio*[6]. Chris Brennan, a compliance manager from the MED Radio Spectrum Management Group, confirmed to the reporter that the terms of the LPFM licence clearly stated there was no interference protection and that two stations could broadcast on exactly the same frequency if they wanted, and that while the ministry encouraged a level of co-ordination among stations they were under no liability to intervene regarding reception problems. Self coordination eventually succeeded in

---

[6] *One Christian Radio* claimed in an earlier article from 8 October 2006 to cover up to 75 per cent of the city of New Plymouth. The urban population of New Plymouth is estimated at 51,000. *One Christian Radio* is therefore possibly broadcasting to a potential audience of over 38,000 people, and in fact quite few of the stations that responded to my survey had potential audiences of around this number.

New Plymouth and the two stations resolved the interference problem amicably.

This story initially came to my attention through the *LPFM user group* on Yahoo! Groups. In the discussion that ensued, many expressed that until such time as the MED modified the GURL to include some mandatory requirements, situations like the New Plymouth case would become the norm, implying that in more crowded environments the so-called "tragedy of the commons" would ensue (Hardin, 1968). Recently the New Zealand LPFM Society, a non-profit group that meets periodically in Auckland has begun accepting and reviewing interference complaints which it then publishes on its website as a type of "name and shame" campaign[7]. A change to the GURL in 2010 has also reduced the former 300 kHz spacing to 100 kHz, effectively allowing more stations to broadcast in the band. As urban and suburban populations become denser and with the ease of acquiring readymade and cheaply imported LPFM transmitters, more and more communities and groups will acquire the tools, skills and individuals to create LPFM broadcasts. The commons is increasingly becoming more crowded but it has, so far, also proven fairly effective at governing itself.

Naom and Benkler have identified two approaches to spectrum management that often materialise: "open access" (Noam, 1998) and "common property" (Benkler, 1998). Open access describes a regime where anyone has access to a specified block of spectrum and nobody controls or limits such access, although payment of an entrance fee may be exacted to self-regulate congestion. Common property regimes promote spectrum as a common pool to be managed by users in a cooperative fashion, much like fish stocks or common pastures, and are legitimised by governments which define their boundaries but play minimal roles in governance or in managing disputes. The GURL creates the conditions for the latter common property regime. Wattage is limited to try to lower congestion and interference, a specific range of frequencies is designated for use, and user management is encouraged.

Critics of the commons will inevitably refer to Garrett Hardin's seminal 1968 text, the *Tragedy of the Commons,* where Hardin argued that an open-access resource will always ultimately be over-exploited and subsequently degraded due to self-interest. While true commons are rarely

---

[7] New Zealand LPFM Society: http://www.lpfmnz.com/.

observed in practice, Benoît Freyens has argued for a more granular approach to describing spectrum regimes and proposes that a wider vocabulary will assist "reform-minded countries" with a "near continuum of policy options" (Freyens, 2009). He argues that while government control is an inescapable feature of any spectrum regime it is the degree to which intervention limits the commons that is valuable.

The limited scope of the LPFM commons may eventually result in two pressures from users and creators upon future spectrum allocation policy. On the one hand, commons are often sites of innovation (Freyens, 2009) and the GURL has created a truly local space in the otherwise virtually inaccessible New Zealand radio market for access, innovation, and a communicative determination to flourish. On the other, the combination of the GURL's power limitations plus tax-payer supported funding structures based solely on audience numbers impossible to achieve under the GURL, serves to further deter improvement of the economic conditions of most LPFM stations, limiting their potential to provide innovative and challenging content. As some stations grow and become essential parts of their audience's lives and tools of local community voices, it is possible communities will eventually lobby for more rights over greater slices of the New Zealand radio spectrum. It can also be argued that funding agencies need broader definitions of funding criteria to fit the changed landscape of community broadcasting after a decade of the GURL.

Conclusion

Most New Zealand LPFM stations display some sincere desire to engage and serve local communities, whether those communities are geographical like *Waiheke Radio* or audience-focused like *UP FM* in Auckland - a station for the local dance music community. Sixty percent of the stations that I surveyed reported playing some local music and many of them received and played tracks from the monthly "Kiwi Hit Discs", CDs that are collections of the latest New Zealand songs funded by NZOA. Most see themselves as offering an alternative to what is offered by full power radio stations, although there are a few Top 40-type stations, and quite a number of contemporary adult music hobbyist stations that imitate mainstream commercial station playlists and music rotations.

An Australian study, *Community Media Matters: An audience study of the Australian community broadcasting sector*, reported a strong desire from Australian communities for more local media ownership as well as for more diversity of content (Meadows, Forde, Ewart, & Foxwell, 2007). A similar community oriented study in the United States by Torosyan and Munro (2010) looked at the viability of local content in community radio, particularly news content, and found that residents of small towns and rural areas generally listened to a wider variety of radio programmes than respondents who lived in moderately populated or larger cities and suburbs. Torosyan and Munro also report that the following criteria were considered as highly important features of local radio:

- Speed and reliability of emergency information including weather forecasts
- Caring about listeners
- Being a reliable source of local news and information
- Friendly personalities

While the least important features included:

- Covering local high school sports
- Offering contests, prizes and giveaways
- Celebrity-type radio personalities

Their study also reported a majority of listeners demonstrating high levels of satisfaction with their local radio service, with over 80% of respondents reporting that they would greatly miss the stations if they disappeared from the airwaves. These studies show that local, community-based and easily accessed broadcast services can enhance community well-being and provide alternatives to the mainstream media that reflect local lives. My own experience and survey suggests that in New Zealand LPFM stations are providing local and meaningful alternatives to the highly commercial national networked radio stations that dominate the New Zealand radio environment. More work though needs to be done in this area as a general lack of scholarly and reformist approaches have enabled mainstream media lobbyists and funding agencies to ignore LPFM, or worse dismiss it with claims of amateurism and insignificance.[8]

While the New Zealand radio deregulation experiment is nearing its 25th birthday, the creation of LPFM broadcasting and the birth of a new spectrum commons is turning 10. It is a good time to reflect on the

---

[8] The most recent NZ On Air Public Perception Research Quantitative survey undertaken in April 2010 still failed to ask one question about radio, instead emphasising television and music operations.

experimental approaches that the organisation of spectrum has undergone in New Zealand. The auction process and the development of an open access radio commons have been innovative and valuable case studies in managing a spectrum resource that is subject to significant scarcity. Spectrum rights allocated by open auction and a lack of regulation of ownership and content have curtailed the civic and local usage of much of the radio spectrum in New Zealand. LPFM on the other has emerged and one would have to say flourished, with almost no official funding and little commercial impact. One of the questions now facing broadcasting policy makers in New Zealand is whether to enhance this new commons or leave it to the fringes of broadcasting and the "enthusiastic amateurs".

References

Benkler, Y. (1998). Overcoming agoraphobia: building the commons of the digitally networked environment. *Harvard Journal of Law & Technology* 11 (1), 287-399.

Cassandra, P. (9 March 2011). Plan to reconnect residents. *Southland Times*. Retrieved December 10, 2010, from
http://www.stuff.co.nz/southland-times/entertainment/4748543/Plan-to-reconnect-residents.

Drinnen, J. & Niesche, C. (21 July2007). Ironbridge higher bid secures takeover. *The New Zealand Herald*. Retrieved December 10, 2010, from
http://www.nzherald.co.nz/radio-broadcasting/news/article.cfm?c_id=263&objectid=10452901

Duignan, G. & Shanahan, M. W., (2005). The impact of deregulation on New Zealand commercial radio, in Neill, K. & Shanahan, M. W. (eds.), *The Great New Zealand Radio Experiment*. Melbourne: Thomson Dunmore Press. pp. 17-46.

Ewart J., Forde, S., Foxwell, K. & Meadows, M. (2007). *Community Media Matters: An Audience Study of the Australian Community Broadcasting Sector*. Brisbane: Griffith University.

Freyens, B. (2009). A policy spectrum for spectrum economics. *Information Economics and Policy*, 21 (1), 128-144.

Hardin, G. (1968). The tragedy of the commons. *Science*, 162 (1), 124-142.

Joyce, Z. (2008). Cultural sites of the spectrumscape: negotiating global flows in NZ radio broadcasting. Conference paper from *Media: Policies, cultures and futures in the Asia Pacific Region*, Curtin University, Australia, 27-29 November 2008. Retrieved December 10,

2010, from
http://mediaasiaconference.humanities.curtin.edu.au/pdf/Zita%20Joyce
.pdf

Maharey, S. (2001). Values and politics: Some reflections on the new
social democracy in a New Zealand context. Retrieved June 8, 2009,
from http://www.beehive.govt.nz/speech/

Mueller, M. (1993). 'New Zealand's revolution in spectrum management'.
Information Economics and Policy, 5 (1), 159-177.

Noam, E. (1995). Spectrum auctions: yesterday's heresy, today's
orthodoxy, tomorrow's anachronism. Taking the next step to open
spectrum access. *Journal of Law and Economics*, 41 (1), 765-790.

Reilly, B. (2011). New Zealand commercial radio – Coke or L&P? Coke
the global brand, or L&P, the local within the Coke empire: The
relationship between America and hybrid culture. Unpublished
conference presentation from *The Radio Conference 2011: A
Transnational Forum*, Auckland University of Technology, Auckland,
New Zealand, 11-14 January 2011.

Rosenberg, B. & Mollgaard, M. (2010). Who owns radio in New Zealand.
*Communication Journal of New Zealand / He Kohinga Korero, Special
Edition: Radio: Challenges and new directions.* 11 (1), 85-107.

Torosyan, G. & Munro, C. (2010). EARwitness testimony: applying
listener perspectives to developing a working concept of "localism" in
broadcast radio, *Journal of Radio & Audio Media*, 17 (1), 33-47.

Tune in, turn on, get miffed, (14 October 2008). Taranaki Daily News.
Retrieved September 10, 2010, from
http://www.stuff.co.nz/taranaki-daily-news/671444/Tune-in-turn-on-
get-miffed.

CHAPTER TWELVE

RADIO AS A TOOL FOR REHABILITATION AND SOCIAL INCLUSION

MATT GRIMES AND SIOBHAN STEVENSON

The following chapter comes from an amalgamation of empirical research, reflection on practice and a desire to find practical examples to support some of the theoretical frameworks within which we work. In this chapter we will discuss ways in which community radio can support the rehabilitation, personal development and inclusion of marginalised individuals and their (re)engagement with mainstream society. Drawing on ideas of policy and practice, we will demonstrate how radio producers and trainers can facilitate personal development, while reconnecting people with society in a meaningful way. In the first part of this chapter we will discuss issues around rehabilitation within the UK penal system, followed by testimonials outlining specific examples to demonstrate how prison radio is being used as a tool to support and address some of these issues. The second part of the chapter will discuss some of the complexities of exclusion and give practice based examples of how radio training has been used; firstly within a specific Gypsy, Roma and Traveller[1] community, and secondly with a group of young people living in a remote rural area, to challenge and re-address some of those issues.

Context

There are many definitions of community radio and international regulators set out their priorities with varying degrees of emphasis, but as this chapter deals with case studies based in the UK, we will refer to the definition laid out by Ofcom[2] the UK broadcast regulator. Ofcom defines a

---

[1] In this chapter the terms Gypsy, Roma and Traveller refer to the whole nomadic community in the UK
[2] http://www.ofcom.org

community as people who live, work or undergo education or training in a particular area or locality, or people who have one or more interests or characteristics in common. It states key characteristics of community radio (such as fostering social rather than commercial gain), can positively benefit members of particular communities. Charles (2010) defines "social gain" as a buzzword used to distinguish community radio and as leverage for gaining spectrum space. Ofcom states that community radio's intention should be to serve one or more communities, and the service should offer members of the target community opportunity to participate in the operation and management of the service. Though the case studies outlined in this chapter are not focused on considerations of regulatory agenda (social gain) they do demonstrate how training in radio production skills and ICT/digital technologies can benefit the recipients' own and wider communities.

Past European projects[3] have outlined community radio as a tool for empowerment, and Lewis and Jones (2006) claim it can help build individual's confidence and make people believe that their lives and the world around them can change for the better. Günnel argues that for community radio and all organisations addressing socially disadvantaged groups "media competence and cultural empowerment are important issues" (2008, p. 87). She goes on to suggest community radio's role is to "give a voice to the voiceless" (ibid). Using examples of past media projects we will demonstrate how some of these ideas and theories intersect with our own media practice.

Though there are many definitions of community radio, most academics discussing case studies refer to the physical space (the station). There are also discussions of the participatory nature of radio, the importance of volunteers and how involvement of individual community members can benefit the wider community (Meadows, 2007; Coyer, 2006; Van Vuuren, 2006). However, as online models demonstrate, the benefits of community radio are not always rooted in the physical space it occupies, or the collective benefits to the community. Instead, the emotional support gained through individual participation of members of online communities in sharing how they feel, can build and strengthen communities (Thompson, 2008). With this in mind, we would argue that community radio operates in an active third space outside of physicality or location, and is more about the development of individuals than communities. In simple terms, Bhaba (1994) outlines third space theory as a virtual place

---

[3] http://www.crosstalk-online.de

where cultures meet. Many academics (Gutièrrez et al, 1999; Millard, 2003; Moje et al, 2004) have used this concept to understand what happens when popular culture meets pedagogy and the hybridity that comes from the merging of these two areas of interest. In our case, the first space is the physical space of the station, the second space is the community of people it serves, and the third space is one in which individuals develop through the acquisition of skills that can then be transferred and used in different community settings. This also suggests that whether attempting to mobilise or strengthen an established community or build and sustain a new one, the key to success is in the development of the individual as this is ultimately beneficial to the wider community.

Siemering discusses community radio as a "catalyst for building communities" (2000, p. 373). This is echoed by Meadows who outlines community radio as a tool for building sustainable communities through "fostering citizen participation in public life" (Meadows et al, 2007, p. 14). Therefore, if we apply these theories to a prison environment this would suggest a radio service has the potential to build a community and foster participation within prison life. However, we must first consider the role of community radio in rehabilitation.

Radio for rehabilitation (Siobhan Stevenson)

Rehabilitation is a complex term, as are the terms inclusion and exclusion, so it is important that we are able to distinguish between meanings in context. Rehabilitation in the context of prisons refers to the time between being convicted of a crime and the conviction being considered "spent".[4] Though this is the subject of ongoing review[5], we will be considering rehabilitation in terms of the health and well-being of individuals. Davis describes rehabilitation as "a generally active, dynamic continual process concerned with physical, social and psychological aspects" (2006, p. 8).

There are ongoing debates around the role of prisons and whether rehabilitation is one of their many responsibilities, particularly as the constant struggle to operate effectively within their budget often means cutting services. However, within most prisons, offenders are encouraged

---

[4] http://www.legislation.gov.uk/ukpga/1974/53
[5] Home Office. (2002). Breaking the Circle: Rehabilitation of Offenders Act 1974 Review. London: HMSO.

to undertake educational qualifications and skills development courses, in an attempt to assist with this rehabilitation process. There is currently a closed broadcast community radio service operating within prisons in the UK, under charitable status and supported by the National Offender Managment Service (NOMS)[6], which is responsible for coordinating a number of services (including radio training) which contribute to reducing re-offending. The National Prison Radio Service (NPRS)[7] initially began broadcasting to the inmates of Brixton Prison and is currently being developed into a national network of stations under the guidance of the Prison Radio Association (PRA).[8] The overarching aim of the service is to broadcast programming specifically aimed at the prison community, but also to teach radio production skills to select inmates, with eligibility restricted to those serving time for non-violent offences. Günnel states that "media competence and cultural empowerment are important issues for socially disadvantaged groups" (2008, p. 87). External agencies such as UNLOCK[9], the National Association of Reformed Offenders, have supported this argument claiming that prison radio is vital because:

> It equips prisoners with employable skills for their release-crucial to reducing re-offending... and it provides a refreshingly open medium for the distribution of information in an environment where knowledge is still largely seen as power and so the privilege of those in authority (Mark Leech, PRA online).[10]

The PRA outlines the radio service as having three main objectives, which include benefitting individual prisoners, the wider prison community and society as a whole (Prison Radio Association 2010). In an attempt to substantiate how these objectives are achieved, we interviewed a former station manager, who at the time of interview, was no longer working in prison radio, but had previously been responsible for one of the prison radio stations. Due to the sensitive nature of the information discussed within the interview and their position within the organisation, they wished to remain anonymous.

When asked if the service fulfils its objectives they answered:

---

[6] http://www.hmprisonservice.gov.uk/abouttheservice/noms/
[7] http://www.hmprisonservice.gov.uk/news/index.asp?id=9452,22,6,22,0,0
[8] http://www.prisonradioassociation.org/
[9] http://www.unlock.org.uk
[10] http://www.prisonradioassociation.org/?con=whypra

It is difficult to evaluate as it is anecdotal. For those listening to and those working on the station, the benefits and experience are different. A lot of prisoners give clear examples of how they have benefited as individuals through listening to the weekly broadcast of the Prison Induction Handbook, which is especially useful for hard to reach prisoners such as those who don't read or mix very well. The prison staff were split, some were for and some against. Those who got it found they could use the station for their benefit to get across basic information which made their job easier and the general environment better (Station manager, personal communication, 2010).

This comment suggests this service does benefit the individual by keeping them and the wider prison community informed, fulfilling the service's objectives and indirectly making the prison staff's jobs easier. However this gives no sense of empowering prisoners and raises questions around how empowerment is possible when freedom and movement are limited in a prison environment. The interviewee suggested this comes through the development of transferable skills.

Through scriptwriting they learn literacy, focus and concentration. Timing your programmes reinforces numeracy and understanding of time, and working to deadlines helps with timekeeping. Whilst making programmes together they learn the value of teamwork, radio production develops organisational skills and presentation skills help them learn to speak with authority. They've made something as well so it gives them a sense of achievement (Station manager, personal communication, 2010).

We argue that this comment should be considered as evidence that, with Davis's (2006) definition in mind, the process of skill acquisition is an important part of the process of rehabilitation. Through the supply of information and development of radio production and transferable skills, individual prisoners are empowered and given a voice, albeit in a closed community setting. However, whether the development of these skills can be transferred to outside communities through the individual, or whether the prison community can be sustained after individuals finish their sentence and leave is not clear, as the interviewee outlines below:

When it came to getting people to a stage where they can apply for jobs, there is not as much help or as many support systems as you would like, but this comes down to funding. Most of the prisoners I worked with have either left prison, in which case there are issues with keeping in touch with them, and others are still serving their sentences, so we won't know until follow up work is done what they are doing now and how it benefited them (Station manager, personal communication, 2010).

Wilkinson and Davidson's (2008) evaluation of the initial West Midlands Prison Radio Taster Project, shows that twenty five out of twenty nine beneficiaries completed the course, gaining an NCFE qualification. Eleven of those moved into further education, training and other government programmes, two moved into full time employment and one became self-employed. At the time of publication of the report, the researchers were still awaiting follow-up data on another fourteen candidates.

Prison radio operates in a closed space with predominantly physical, as opposed to cultural boundaries. However, previous projects have attempted to work with marginalised individuals restricted by cultural and other social boundaries. Going back to ideas of an active third space, the second part of this chapter will now draw upon reflections of radio practice that were employed to assist with and support the re-engagement of two specific groups of people who are often perceived as excluded.

Radio for social inclusion (Matt Grimes)

The following study gives a reflective account of media based practice, drawing upon two previous community based radio production projects that were developed and delivered between 2006 and 2007 with two different groups of people that could be perceived as socially excluded or marginalised. This study will draw on research by academics and practitioners to contextualise and support some of the issues raised, engaged with, or discovered while delivering these projects. This is also supported by verbal and written feedback obtained from the participants during the evaluation process of the projects. The purpose of the projects was to discover if and how access to and engagement with digital media technologies could help in the processes of social inclusion for these particular groups. The first project was conducted with a small group of Travellers and the second with a group of young people living in rural communities.

Social inclusion can be seen as the development and establishment of positive and affirmative actions by organisations, groups of people and individuals to help address or change some of the factors that lead to the social exclusion of others within a particular society. It's opposite, social exclusion, is a well documented and debated subject (Collins, 2003; Farrington & Farrington, 2005; Labonte, 2004; Orr, 2005; Shucksmith, 2004; Silver, 1994; Young, 2000). The notion of exclusion encompasses many determining factors including race, culture, health, gender, social positioning, wealth, employment and education. In the face of this diversity of factors, then, social exclusion is a mutable term that is open to numerous definitions and interpretations. Silver discusses in detail some of the problematic issues that arise when trying to define social exclusion, as it is "loaded with numerous economic, social, political and cultural connotations and dimensions" (1994, p. 535). This is further reiterated by Orr who writes:

> Nevertheless, a theme common to most, if not all, definitions of social exclusion is that social exclusion is multidimensional. That is, often, social exclusion will have multiple necessary conditions. So, it is usually the combination of factors that cause social exclusion (Orr, 2005, p. 4).

Social exclusion is often also associated with marginalisation. Young (2000) argues that it involves and impacts on those who are excluded from policies or prevented from accessing services and programmes. For the purpose of this chapter, we will employ the notion that social inclusion is about people's access to resources and opportunities that are readily and otherwise available to members of society, to enable them to fully participate in and socially integrate with society. Access to specific communicative resources is closely linked to the development of media literacy and its effectiveness in allowing participation in society and the economy. These issues were highlighted in an audit on media literacy conducted by Ofcom, the UK's independent broadcasting regulator, who were given the responsibility for promoting media literacy in the UK as part of new legislation contained within the Communications Act 2003.[11] This is further supported by a media literacy dissemination project[12], conducted in 2009 by the Community Media Association[13] which focused on the importance of access and understanding of media. Their findings also highlighted issues around digital exclusion and how some groups within society were at risk of being left behind in terms of ownership of new technologies and the confidence and competence to use them. Members of these groups were also identified as at risk of being culturally, politically, socially and economically excluded.

Castells' (2001) study of the Internet and digital technologies suggests that not only do digital technologies have the ability to liberate dominated

---

[11] http://www.legislation.gov.uk/ukpga/2003/21/contents

[12] http://www.commedia.org.uk/2009/05/14/your-media-your-tools-update/

[13] http://www.commedia.org.uk/

groups but also to marginalise and exclude those who have no or limited access to those technologies. Frediani rightly notes that "The question of access brings us to the issue of the digital divide and marginalised minorities" (2010, p. 257) and cites some key studies in the area of the "digital divide" and its impact on minority engagement with digital technologies (Mehra, Merkel & Bishop, 2004; Miller & Slater, 2003). Frediani also argues that the Internet and other information and communication technologies (ICTs) are widely considered as instruments of empowerment, which supports Günnel's (2008) arguments about media competence leading to cultural empowerment. In parallel it could be argued that access to ICTs, such as portable digital recorders and laptop computers enabled with mobile Internet access, could be employed to help liberate marginalised and socially excluded groups.

Method and practice I

In 2005 I was invited to join the editorial board of the Travellers Times[14], a quarterly magazine that is written for and by the Gypsy, Roma and Traveller communities in the UK. Historically, Traveller communities have been demonised and excluded by Western society and the media (McCann, 1994; Puxon, 1987; Ryder, 2002). This in turn has established, particularly in Europe, a misrepresentation of both them and their culture that is, at times, further reinforced by the media. Frediani's (2010) study on the Internet and Gypsy/Traveller activism points to the fact that for many travelling communities, who have limited access to information and communication technologies (ICTs), there is little recourse available to them to challenge and address these prejudices and misrepresentations in society and the mainstream media. Through my own experience of being a member of a travelling community, and from engaging with other Travellers, I knew that some of these earlier observations regarding the limitations of access to digital ICTs and media misrepresentations could be corroborated. Some ICTs, such as computers, the internet, mobile internet, high-speed data networks, digital radio and television for remote learning, had not reached some of these communities, which led me to consider how they may be used to challenge and address some of the issues around exclusion and misrepresentation.

---

[14] http://www.travellerstimes.org.uk

In early 2006 I designed and delivered a pilot radio production course for a group of Travellers in Herefordshire[15] specifically tailored to the needs of that particular travelling community. The course was not commissioned by any organisation, but instigated by myself as a way of developing my own media training practice. I also felt that this particular group could possibly benefit from a project of this sort as it was anticipated that they would be able to use digital ICTs to document their culture and therefore have a means of challenging and addressing some of the media and cultural misrepresentations. The course was specifically designed to fulfil two central goals that might help in that process. The first was to enable and train volunteers within that specific community to engage with basic practices of radio and media production, allowing them to produce and edit their own audio material. The second was to enable the participants to then train other Traveller communities to do likewise, thus extending that skill base among a wider community. This pedagogical process of cascading skills and knowledge to be passed on to other members of their community was to some extent imitating and replicating the familiar oral traditions of Travellers.

On reflection, some of the methodologies employed in this and the other project were similar to those discussed by Günnel in *Action Oriented Media Pedagogy* (2006). These she outlines in her article "The 'Dual Role' approach: Encouraging access to community radio" stating that:

> Teaching using the action oriented media pedagogy approach should:
> *   be based upon and refer to the learners' life-reality
> *   be structured so as to activate the participants
> *   take a learner-centred approach
> *   support learning through action
> *   be product-oriented (Günnel, 2008, p. 89)

Günnel states the dual role approach requires the participants to be:

> ...at one and the same time acting as trainees acquiring new skills and understanding, but also imagining themselves in their role as trainers as they absorb the methods and style of the course and think of ways they can adapt and present the material in the future (Günnel, 2008, p. 89).

The radio project for the Travellers operated for a period of five days (thirty hours) involving four participants of mixed gender and ages ranging

---

[15] The participants wished to remain anonymous.

from nineteen to twenty five years of age. The course was designed to be progressive and cumulative, where each day's workshop covered a new subject area that built on the previous session. Communication and interviewing skills were developed through role playing scenarios where the participants would alternate between being interviewer and interviewee to encourage and develop their confidence in talking comfortably within a group and to enable a greater understanding of the skills required to produce an informative and coherent interview.

> I really liked the part where we had to pretend to be a celebrity and answer questions from the interviewer, it was a laugh and made me feel better about talking in front of people because it was done in a fun way. When I had to be the interviewer it was harder though 'cos [sic] I had to come up with some interesting questions to ask but it did make me think more about what I would ask to make my interviews more interesting (Female Traveller, aged 19).

Basic radio production and digital audio recording and editing skills were taught through the process of demonstrations and practice using hand held digital recorders and basic editing software on laptop computers. There were initial concerns, among the participants, about learning how to use the equipment, but the ease and speed at which the participants acquired and executed their new found skills confirmed what I had experienced on other similar projects, in that it helped develop their confidence when engaging with digital technologies.

> I thought you had to be really clever to learn that stuff we did with the recorders and stuff but it was easy once I done it a few times (Female Traveller, aged 22).

> The computer editing was really hard at the beginning and I never believed I would ever understand it. But now I can sort of do it I just want to get better and better at it (Male Traveller, aged 19).

Once the participants felt confident in using the equipment, they had to take on the role of the trainer and teach the necessary skills back to the project leader/trainer and to each other. The purpose of this was to reinforce those skills and also demonstrate that they were proficient in teaching those skills as one of the central outcomes of the project was to 'cascade' that knowledge of production skills to other members of their community. Their newfound "media competency" (Günnel, 2008, p. 87) gave them the basic skills to enable them to produce and document media artefacts about their own cultural group that could be shared among each other and with a wider audience. This type of project could be seen as an intervention into addressing some of the issues raised by both Ofcom and the Community Media Association regarding the need for media literacy development among marginalised and excluded groups. In Ciolek's study of the usefulness of the Internet he states that members of marginalised and excluded groups can:

> ... use the Internet to: intensively liaise and network amongst themselves and with other friendly groups; document their culture, language, history and achievements; inform and educate the neutral sections of public opinion about their plight and grievances (Ciolek, 2004; cited by Frediani, 2010, p. 267).

It could be argued that the use of digital ICTs to further develop media competence and content among excluded groups can serve a similar purpose. The consideration of the specific cultural and learning needs of this group, and the incorporation of those into the design of the learning programme, contributed to making the project relevant and focused in relation to the pre-determined project outcomes. The use of a pedagogical approach that was similar to Gunnel's "dual role" approach helped the participants in the project achieve the two central learning outcomes; first, the participants learning new skills, and second, the ability to share those skills among their community. It could also be argued that this created an opportunity for this community to start developing media artefacts which could challenge and address some of the misrepresentations levelled at them.

Method and practice II

Referring to the definition of community laid out by Ofcom and discussed earlier in this chapter, I was also interested to see whether the engagement with and use of digital ICTs, specifically digital recorders and editors, could support the construction and development of a *new*, more cohesive community from within an existing sporadic one, in contrast to aiding an *established* community such as the Travellers. At the invitation of a youth worker from a rural Herefordshire youth club, a second project was structured and delivered to a group of young people living on remote hill farms on the English/Welsh borders. This particular group of young people felt socially isolated and excluded due to their geographical location and particularly their limited access to transport and leisure.

> The nearest bus stop is 2 miles from our farm and there's only 4 buses a day so if I want to go anywhere at the weekends or in the evening I have to ask my mum or dad and if they are busy it means I don't go (Female participant, aged 15).

> The last bus from Hereford city centre to my nearest village is 6.30 in the evening which mostly means I can't meet up with my friends to go to the cinema or whatever. At times I feel my social life is non-existent cos' [sic] I'm stuck at home when all my mates are in town hanging out and having a laugh (Male participant 17) .

Shucksmith (2004) notes that issues such as access to transport and leisure contribute to people's perceptions and experiences of social exclusion. The rationale for the project was to enable young people in a specific rural area to have access to and engage with media and digital technologies, and to assess how the technologies could be utilised as a cohesive communication tool that would support the limited impact that sporadic youth club meetings had. A project was specifically designed for this particular group of eight teenagers after an initial consultation meeting with them to determine what outcomes they would want to get from the project. They wanted to be able to engage with new digital media, learn new practical and communication skills, have the ability to document and share their experiences, and have a greater understanding of how radio features are produced. Similar to the previous project with the Travellers, the course was designed to be progressive and cumulative, where each day's workshop covered a new subject area that built on the previous session.

Communication and interviewing skills were developed through role playing scenarios where the participants would alternate between being interviewer and interviewee, to encourage and develop their confidence in talking comfortably within a group and to enable a greater understanding of the skills required to produce an informative and coherent interview. There were fewer concerns from the participants about learning to use the equipment, as they had all developed reasonably good ICT skills in school. Each week the participants would develop a new production and/or communication skill using basic digital recording and editing equipment and training resources. They had the opportunity to take the equipment home with them each week and independently develop these skills. Between each workshop, directed tasks such as recording in various locations and digital editing exercises would be set, and then assessed, evaluated and discussed by the group at the beginning of each following workshop. The peer review and evaluation sessions allowed the trainer to monitor individuals' progress and identify specific weaknesses, allow for peer and self critique of participants productions, and encourage the participants to develop and improve their work.

> It became quite competitive between us all seeing who could come up with the most interesting idea or the best recording or smoothest editing. It was really exciting waiting to hear what the others in the group had done and who got voted the best of the week. It really made me think much more about what I was going to do and how (Male participant, aged 17).

The project structure provided the participants with the necessary basic production and communication skills, which gave these young people the opportunity to express themselves creatively. By self-producing audio diaries, short documentaries and audio features they were, as developing teenagers, able to share their experiences, aspirations, hopes and fears with each other by uploading their audio material onto a purpose built project website. The website was designed not only as a hub for the participants to share their material but also for them to communicate with each other, outside of the sporadic youth club meetings, and to enable their experiences to be shared with a wider audience. Participants' feedback identified many benefits to the project including helping build their confidence by developing new communication and technical skills; personally contributing to something that was tangible and had positive value; and being able to express themselves, their experiences, their feelings and issues that personally affected them, and in doing so helped reduce their feelings of isolation.

> I really enjoyed learning to make the recordings. At the start when we first got to use the recorders it was weird and a bit embarrassing hearing back my own voice, but once I had gotten used to it, it was fun. I never thought I would but I really liked talking about what I do and stuff, and thinking that other people might be able to listen to it to and maybe understand what it is like being a teenager growing up and living in the countryside (Female participant, aged 15).

> I like being able to share my ideas and the things I think about with other teenagers in the same situation as me and knowing that I'm not the only one feeling lonely and bored and having no one of my age living close by to hang out with.(Male participant, aged 16).

As a result of the positive experience they had, the young people, along with some help and support from the youth service and the project team, took ownership of the project after the programme's cessation. They identified pockets of available youth funding, both locally and regionally, and set up their own small Internet radio station, streamed through the project website. This online platform enabled them to keep in touch with each other and establish new relationships with young people in other rural areas via broadcasts, podcasts, forums and embedded social networking platforms.

> I use MSN messenger and Facebook a lot to chat to my friends from 6th form (college) but this is like different 'cos [sic] with these friends I feel we are like part of a special group that have come together to do something different. It's brilliant 'cos [sic] I have a new set of friends from doing this project and by using the (website) forum I now have other new friends not only from nearby but from other places as well who are similar to me, living on farms in the middle of nowhere with nothing to do (Male participant, aged 17).

The inclusion of an online platform for this group reiterates some of the ideas that Thompson (2008) discusses, about how communities can be built and strengthened by the individual participants sharing their feelings and gaining emotional support from others of that online community. One of the interesting and central issues here is that not only can simple radio production skills support communities, as was demonstrated in the project with the group of Travellers, but in this particular case it contributed significantly towards constructing new and vibrant community. Their engagement with and use of radio production techniques, combined with other digital technologies and ICTs, developed skills within the group that enabled them to communicate in a more meaningful way than the previously sporadic youth club meetings. This, in turn, enabled the group to feel more connected and, through the development of their website and its content, extend the notion of the community beyond its original geographical locality.

Conclusion

The purpose of this chapter was to discuss ways in which community centred radio can support the rehabilitation, personal development and inclusion of marginalised individuals and groups, and their (re)engagement with mainstream society in a meaningful way. In doing so we drew on ideas of policy and practice and used specific case studies to demonstrate how radio producers and trainers can facilitate personal development and inclusion of marginalised individuals and groups. This was demonstrated by investigating three different communities who have restricted access to resources and opportunities, due to physical, cultural and geographical barriers. In doing so we found that access to resources and opportunities, specifically digital technologies and ICTs, contributed to the development of the individual and subsequently their community. The acquisition of these new skills could be transferred to other social, economic and community settings, dispelling myths and challenging ideas around exclusion and marginalisation.

In the prison radio project it was found that prison radio not only serves people through skills acquisition but through providing basic information in broadcast form. Through learning radio production skills, inmates also developed softer transferable skills in what we argue is an active third space (where they also develop confidence and self esteem). These skills could be used in other settings and contribute to re-engaging with mainstream society. Similarly the projects with the Travellers and young rural people also highlighted the benefits of how acquiring skills contributes to the personal development of the individual and in turn benefits the wider community.

We also concluded that programmes of learning need to be tailored to the specific needs of the individual or group, as we identified that the learning outcomes of each specific group differed, as did their circumstances. In the project with the young rural people, the peer review and evaluation sessions proved extremely useful not only to the trainer, as a monitoring process, but also to the participants who commented on the competitive value gained from this pedagogical approach. Whereas in the project with the Travellers a pedagogical approach similar to that of Günnel's (2008) dual role approach helped to reinforce their new skills while they developed the means to cascade those skills to their peers in a way that was culturally receptive. The value of this research is in validating through practical examples the importance of media competence and cultural empowerment, and how these pedagogical approaches allow people to take charge of their own voice (giving a voice to the voiceless). This is empowering because it allows people to be part of constructing and sharing their own stories, potentially dispelling myths and misrepresentation, which could in turn lead to recognition and respect.

Another interesting issue here is that not only can simple radio production skills support communities, but in the case of the young rural people it contributed significantly towards constructing and sustaining a new community that was more cohesive than the original and sporadic community it emerged from. Their engagement with radio production techniques enabled the group to feel more connected and extended the notion of the community beyond its original geographical locality.

There are numerous difficulties surrounding the evaluation of future outcomes of reflective media practice, especially where future evaluations for research purposes were not within the original project remit. Some of those difficulties include current access to the specific participants and respondents. The individual beneficiaries of the prison radio training have the right to privacy and anonymity upon release. With the Travellers, their particular nomadic lifestyle has made it hard for the authors to locate and contact them. As a result of this, it has not been possible to assess whether the cascading form of pedagogy and the pedagogical approach similar to Günnel's dual role methodology was successful in terms of challenging some of the misrepresentations of their culture. The project with the young rural people was conducted within the local community of one of the authors, who intermittently monitored its efficacy, until the group disbanded after eighteen months due to loss of funding and changing personal circumstances of some of the participants.

Though we cannot conclude that the groups in the case studies are any less marginalised or excluded as a result of the projects, we would suggest that the engagement with digital technologies and the acquisition of skills presented them with access and opportunities for challenging some of the key issues around rehabilitation and inclusion discussed respectively by Davis (2006) and Young (2000). Similarly, these projects could be seen as an intervention into addressing some of the issues raised by both Ofcom and the Community Media Association and the need for media literacy development among marginalised and excluded groups.

References

Bhabha, H. K. (1994). *The Location of culture*. London: Routledge.

Castells, M. (2001). *The Internet Galaxy*. Oxford: Oxford University Press.

Charles, H. (2010). *UK Community Radio: Policy Frames & Outcomes*. London: London School of Economics & Political Science, Department of Media & Communication

Ciolek, T. M. (2004). Internet and minorities. In C. Skutsch, (Ed.), *The Encyclopedia of World Minorities*. London: Routledge.

Collins, H. (2003). Discrimination, equality and social inclusion. *The Modern Law Review, 66*(1), 16-43, doi: 10.1111/1468-2230.6601002

Coyer, K. (2006) *Community Radio Licensing and Policy. An Overview.* Goldsmiths College. University of London

Davis, S. (2006). *Rehabilitation: The use of Theories and Models in Practice.* London: Elsevier Ltd.

Farrington, J., & Farrington, C. (2005). Rural accessability, social inclusion and social justice: Towards conceptualisation. *Journal of Transport Geography , 13*(1), 1-12,
doi:10.1016/j.jtrangeo.2004.10.002

Frediani, M. (2010). The web against discrimination? Internet and Gypsies/Travellers activism in Britain. In M. Stewart, & M. Rovid (Eds.), *Multi-disciplinary Approaches to Romany Studies: Selected Papers from Participants of Central European University's Summer Courses 2007-2009* (pp. 257-270).

Gutiérrez, K. D., Baquedano-López, P. & Tejeda, C. (1999). Rethinking diversity: Hybridity and hybrid language practices in the third space. *Mind, Culture, and Activity, 6*(4), 286-303,
doi: 10.1080/10749039909524733

Günnel, T. (2006). Action-oriented Media pedagogy: theory and practice. In P. Lewis, & S. Jones (Eds.), *From the Margins to the Cutting Edge: Community Media and Empowerment* (pp. 41-65). Creskill, NJ: Hampton Press.

—. (2008). The 'Dual Role' approach: encouraging access to community radio. *The Radio Journal-International Studies in Broadcast and Audio Media , 6*(2 & 3), 87-94, doi: 10.1386/rajo.6.2&3.87/4

Labonte, R. (2004). Social inclusion/exclusion: dancing the dialectic. *Health Promotions International , 19*(1), 115-121,
doi: 10.1093/heapro/dah112

Lewis, P. M., & Jones, S. (2006). *From the Margins to the Cutting Edge: Community Media and Empowerment.* Cresskill, N.J: Hampton Press.

McCann, M.; O'Siochain,S; & Ruane, J. (1994). *Irish Travellers: Culture and Ethnicity.* Belfast: Institute for Irish Studies.

Ewart J., Forde, S., Foxwell, K. & Meadows, M. (2007). *Community Media Matters: An Audience Study of the Australian Community Broadcasting Sector.* Brisbane: Griffith University.

Mehra, B., Merkel, C., & Bishop, A. P. (2004). The internet for empowerment of minority and marginalized users. *New Media Society 6*(6), 781-802, doi:10.1177/146144804047513

Millard, E. (2003). Towards a literacy of fusion: new times, new teaching and learning? *Literacy 37*(1), 3-8, doi: 10.1111/1467-9345.3701002

Miller, D., & Slater, D. (2003). *The Internet: An Ethnographic Approach.* Oxford and New York: Berg.

Moje, E., Ciechanowski, K., Kramer, K., Ellis, L., Carrillo, R., & Collazo, T. (2004). Working toward third space in content area literacy: an examination of everyday funds of knowledge and discourse. *Reading Research Quarterly,* 39(1), 38–70, *doi: 10.1598/RRQ.39.1.4*

Orr, S. W. (2005). *Social Exclusion and the Theory of Equality:The Priority of Welfare and Fairness in Policy.* London: University College London.

Puxon, G. (1987). *Roma: Europe's Gypsies.* London: Minority Rights Group.

Ryder, A. (2002). The Gypsies and exclusion. *Social Work in Europe ,* 9(3), 52-60.

Shucksmith, M. (2004). Young people and social exclusion in rural areas. *Sociologia Ruralis , 44*(1), 43-59, doi: 10.1111/j.1467-9523.2004.00261.x

Siemering, W. (2000). Radio, democracy and development: evolving models of community radio. *Journal of Radio Studies, 7*(2), 373-378, doi: 10.1207/s15506843jrs0702_10

Silver, H. (1994). Social exlusion and social solidarity: Three paradigms. *International Labour Review ,*133 (5, 6), 531-578.

Thompson, K. (2008). *Making virtual communities.* Retrieved April 17,2011, from http://www.bioteams.com/2008/08/06/making_virtual_communities.html

Van Vuuren, K. (2001). Beyond the studio: A case study of community radio and social capital. *The Australian Community Broadcasting Series.* Retrieved December 12, 2010, from http://www.cbonline.org.au

—. (2006) Community Broadcasting and the Enclosure of the Public Sphere. *Media, Culture & Society* 28, 379-392 .

Wilkinson, K. & Davidson, J. (2008). Evaluation of the Prison Radio Association's Activity. Retrieved December 12, 2010, from http://www.prisonradioassociation.org/library/documents/PRA_Exec_Summary.pdf

Young, I. (2000). Five faces of oppression. In Adams, M., Blumenfeld, W.J., Castañeda, R., Hackman, H. W., Peters, & Zúñiga, X. (Eds), *Readings for Diversity and Social Justice: An Anthology on Racism, Sexism, Anti-semitism, Hetrosexism, Classism and Ableism.* New York. Routledge.

CHAPTER THIRTEEN

COMMUNITY RADIO AUDIENCE RESEARCH

JANEY GORDON

There are a number of debates as to the merits and validity of conducting formal audience research in the setting of a community radio station. It may be suggested that the small numbers involved may invalidate quantitative audience research at a community station. Furthermore, it might be suggested that the high social value and impact of community radio is not well measured by the same audience "number crunching" techniques relied on by larger commercial radio stations. With regard to qualitative research there may be a tendency for small stations to point to the social value gained by the group of volunteer broadcasters, but less attention may be paid to the listeners who do not come into the station premises or take part in its activities in a more face-to-face way. In some parts of the world community radio broadcasters and their audiences are in actual physical danger. The radio stations maybe regarded as operating illegally and without the authorisation of the state, they may be regarded as traitors to a regime that is in itself undemocratic or functioning under a violent dictatorship. The station is possibly the only opposing voice and the volunteer broadcasters see their job as sustaining their community through difficult times. It is clear that in these circumstances to conduct audience research is unrealistic.

However in general, community stations operating in more liberal democratic areas, where community stations function as an additional broadcaster to other mainstream radio and they operate under a legitimate licence to broadcast, community radio stations have little to fear from discovering more about those who value their services. It could be argued that community radio stations in these countries have enhanced their communities' lives by providing cultural pleasures, information and the articulation of community concerns, and in addition, a wider understanding of the media, a "media literacy".

The collection of data in a reliable manner help to demonstrate that the station is doing what it has set out to do, both for its own community and for the licensing body. Research can monitor station initiatives and review the programme content and how this is impacting on the community involved with the station. In addition formal research can track the social impact of the station for the volunteers in a more systematic rather than anecdotal way.

The value of this information is that it can be used by the station to review and enhance its work. It also becomes valuable information with which to approach funding bodies or commercial organisations that wish to know who the audience is. Some organisations value community radio as it reaches groups that may be considered "hard to reach" by other means, for example new migrant groups. In addition, even the most benevolent non-governmental organisation or charity also needs to be able to demonstrate that it is spending its resources wisely and that working with a community radio station can be evidenced as financially wise and sound.

This article suggests two research models that can be used by individual stations to establish a robust and appropriate methodology for examining a community radio station and its relationship with its listeners. It is suggested that both quantitative and qualitative audience research are applicable to the community radio sector. In particular it urges community radio stations to conduct their own audience research using simple techniques, which produce data that other researchers and commentators may have confidence in. The research methods have been used and piloted in the United Kingdom, but it is suggested that they would be applicable and relevant to stations elsewhere.

Literature

To understand why community stations generally may seem reluctant to engage in audience data collection and evaluation it is worth reviewing the earliest literature concerned with community radio. AMARC, The World Association of Community Radio Broadcasters, held its first conference in 1983 and from that it became evident that community radio was a worldwide sector of broadcasting worthy of serious review and research into its attributes and impact. In fact by this time United Nations Educational, Scientific and Cultural Organisation (UNESCO) had already recognised that this was a form of broadcasting worth examination and commissioned a number of studies around the impacts and social potential of community media (Berrigan, 1977, 1981 & Lewis, 1993). However Lewis and Booth (1989) summarised the perceived problem of audience research for community radio by explaining what public service and commercial broadcasters do:

"Whereas public service broadcasting starts off with an idea of what the public needs, and recognises [....] the fact that it is difficult to find out what audiences want, marketing tries to define a target audience [...] Public service broadcasting is in the business of selling programmes [programmes which inform, educate and entertain] to audiences, commercial broadcasting [is] in the business of selling audiences to advertisers." (Lewis and Booth, 1989, p. 99)

Community radio is often attempting a more subtle task, it tries to promote and facilitate volunteers and listeners to suggest and make programming material that would not otherwise be broadcast. Reading the poignant examples of community radio in Bruce Girard's collection of community radio articles (1992), it may be seen that the community radio activists and broadcasters at this time, felt little need to find out more about their audiences. They were often core members of the communities that they were broadcasting to and had rapid feedback from listeners about the programmes and issues broadcast, through personal conversations, letters and phone calls.

Ironically it is in the highly commercial environment of the United States that systematic research into community radio is found during this period. The anthropologist E.B. Eiselein conducted audience research amongst native North American Indians to find out the percentage who listened to their community radio stations in May-June 1992 (in Keith 1995, p. 100). Keith also quotes the managers of native stations that regarded audience research as fundamental to what they were doing. "Surveys are valuable for native stations, because they, more than any other want to be in touch with their audiences ... our surveys have encouraged us to do what we do" (Maria, as quoted in Keith 1995 p. 101).

Anthropologists have found the systematic investigation into community radio a self evident area of research. It is amongst ethnographic anthropologists such as Tacchi et al (2003) and Hearn et al (2009) that effective methodologies for investigating the impact of community media initiatives in developing areas has been tried and tested. However in the past decade, the Australian community media have been the most diligent in conducting organised research and data gathering concerning their sector. For other community media such as the United Kingdom's relatively new community broadcasters, the Australian research methodologies and data have been enormously helpful and relevant. A major report was undertaken and written up by scholars at Griffith University, Brisbane in Queensland and looked at the community broadcasting sector as a whole (Forde et al, 2002). This was followed by a second report (Meadows et al, 2007), which was an in depth qualitative study of the sectors audience using established techniques. The research obtained valid and robust results that were favourable and positive for the sector, illustrating the value of the research as well as the sector of broadcasting. Since 2003, *CBOnline* (http://www.cbonline.org.au) has supported a "Community Broadcasting Database", which collects information annually from Australian community broadcasters on a range of topics. Most importantly, since 2004, The Community Broadcasting Association of Australia have funded a biannual survey from McNair Ingenuity a professional research company, which uses market research techniques to gather data on the Australian community media audiences. These extraordinarily rich resources of both quantitative and qualitative research have suggested that community broadcasters will benefit from audience research.

In the 2009 database the report states that with regard to audience research,

> "Community radio stations are increasingly taking responsibility for developing a better understanding of their audiences. In 2007-08 survey, 46 per cent of all stations reported that they either commissioned or subscribed to audience measurement surveys in their service areas. This was an increase of 12 percentage points on 2005-06 when 34 per cent of stations reported undertaking their own audience research. Stations with higher incomes conducted the most audience research: 64 per cent of stations earning more than $500k annually conducted some form of audience research, while 43 per cent of stations earning under $500,000 surveyed their audience. Stations serving religious interests also reported a higher level of audience surveys, with 61per cent commissioning surveys". (CBOnline, 2009, p. 3)

In the United Kingdom, community radio was piloted in 2002 and two research reports were conducted by Anthony Everitt on behalf of the Radio Authority, the radio regulatory body in the United Kingdom at that time (2003a and 2003b) which was concerned with the feasibility of the sector generally (the Radio Authority was subsumed into the new United Kingdom communications regulator "Ofcom" in 2003). A further study into the audience was conducted in 2004, which conducted quantitative and qualitative research into four contrasting radio stations on air at the time (Research Works, 2004). However these reports were commissioned by the regulatory body to examine the new sector and assess its overall viability, rather than discover details about listener appreciation and engagement.

Methodologies

Two simple methodologies for community radio audience research are detailed here, one each for qualitative and qualitative data. Both forms of research are important for community radio broadcasters. Quantitative data gives confirmation that there are listeners and over several surveys this can demonstrate changes in the general awareness of the station as well as an increase or decrease in the audience following initiatives by the station. However, without qualitative research, this numerical data gives little idea as to what the audience appreciate about the station's outputs. Similarly, qualitative results from interviews with a handful of the listeners may indicate great enthusiasm for some parts of the programming, but give no great feel as to how widespread this appreciation is in the community served. In order to interpret data gained from audience research both quantitative and qualitative data need to be undertaken, although not necessarily simultaneously.

Quantitative research

Between 1986 and 2002 small stations were broadcasting in the United Kingdom for 28 day periods under a temporary licensing scheme, known as the Restricted Service Licence or RSLs (Gordon, 2000). It was important even for these small stations to conduct audience research. In order to maintain the licence for successive years funding was required. Some RSLs had to show parent organisations that what they were doing was valuable in social terms, and frequently these temporary stations were the embryos for longer term commercial or community radio and therefore needed to show that there was an audience for the broadcast service they sought to provide.

University RSL stations were an obvious place to conduct quantitative research and at the University of Bedfordshire in Luton, radio students conduct radio audience research annually. The methodology used is suitable for small scale research projects and viable for a small organisation to complete. The methodology was published in the magazine of the United Kingdom Community Media Association and has been used by other RSLs and community stations (Gordon, 2006, pp. 20-21).

In order to find out how many people are listening to a community station, the information can be based on a relatively small sample group. Professional audience researchers have found that a survey using a sample group of 1000 respondents will give a fairly accurate picture of an audience and that a larger group does not give much more precision. However a group of 100 respondents is regarded as the minimum number for a survey, the difference being in the degree of accuracy. If 1000 people in a station's broadcast area are asked "in the past month have you listened to the community radio station [name of the station]?" and 655 say "yes" the researcher can say with confidence that 65.5 per cent of the sample have listened, but with a reasonable estimated margin for error of +/- 2.5 per cent included, a more reliable figure of between 62 per cent and 68 per cent of the total area have listened. However, if 100 people are surveyed and 65 say they have listened, the researcher can still assume with 95 per cent confidence that 65 per cent of the sample group have listened, but a more variable margin for error of +/- 10 per cent is reasonable for the smaller sample size, so between 55 per cent and 75 per cent of the total station area listened. A commercial media group making financial decisions will tend to use a larger sample to ensure greater accuracy but a community station will find that a sample of 100 respondents is a good indicator of its audience (Gordon, 2006).

Radio audience research at the University of Bedfordshire is conducted by under-graduate students during April and May each year. Ten volunteers are asked to conduct short questionnaires in the street with ten people who are not known to them and who are aged 16 to 29. A tally system on a single side of A4 is used to record the results. The key questions being does the respondent know of the station and have they listened to it within the past four weeks. Additional questions gain insights into the time of day that they listen and allows for other brief comments, as well as confirming their age group.

To convert the percentages into actual numbers, the demographics for the area are needed. In the United Kingdom, these may be found online at www.statistics.gov.uk. Access to similar data in other countries varies, but local government offices, libraries or non-governmental organisations working in the area may have usable demographic data.

Radio LaB 97.1FM is used here as an example to illustrate the usefulness of the methods discussed.

The population of Luton according to the Office of National Statistics midyear estimate for 2009 was 194,300 with around 47,700 people aged 16 to 29. The *Radio LaB 97.1FM* audience survey found that 35 out of the 100 16-29 year olds asked listened to the station during the previous month, this suggests that the station has around 16,700 listeners.

47,700÷100x35=16,695

The margin of error is +/- 10 per cent, so it can be assumed, with 95 per cent confidence that the audience is between 12,000 and 21,500:

47,700÷100x25=11,925
47,700÷100x45=21,465

To reduce the margin it is necessary to survey more respondents. Using a sample of 300 reduces the margin of error to +/- 5 per cent, so if 105 people out of 300 (35 per cent) had listen to the station, it could be suggested that between 14,500 and 19,000 had listened:

47,700÷100x30=14,310
47,700÷100x40=19,080

It must be noted that researchers should use this methodology with caution, acknowledging that the respondents are all from the target audience, not the entire population in the radio station's broadcast catchment area. It should not be assumed that if 35 per cent of young people are listening to the station 35 per cent of the whole area is also listening. It must also be remembered that once the figures drop, they become more unreliable. If 35 people out of 100 have heard the station and 12 of them listen to the breakfast show, it is not viable to conclude that one third of the audience listen to the breakfast show because it is not a large enough group to provide statistical reliability.

Qualitative research

In the United Kingdom community radio stations are required to provide "Social Gain". This is defined as,

> The provision of sound broadcasting services to individuals who are otherwise underserved by such services, the facilitation of discussion and the expression of opinion, the provision (whether by means of programmes included in the service or otherwise) of education or training to individuals not employed by the person providing the service, and the better understanding of the particular community and the strengthening of links within it (Article 2 of the Community Radio Order, 2004).

In other countries there may be different definitions but community radio often attempts to benefit an audience that is underserved, facilitate local discussions, act as a training provider and generally provide a better understanding of the community than might otherwise exist. It may be thought of as providing social worth or Putnam's "social capital" (Putnam, 2000, pp. 22-23). However, these benefits are difficult to evaluate. A listener focus group is a good way to ask listeners direct questions about how they feel about the station and allow them to prompt each other into recalling and sharing examples. It puts a human face on the bare statistics of quantative audience research.

A team at Griffith University in Brisbane, Australia conducted an in depth research study on community radio audiences in Australia (Meadows et al, 2007). Their methodology involved a series of focus groups at a selection of Australian community stations, which were conducted by academic researchers and sought to give an overview of listener's views. The methodology used provides an excellent resource for other community stations wanting to conduct qualitative research of their audiences. A modified version has been tested in the United Kingdom for use by community stations themselves. The adapted methodology has been published online as a part of "Prove It" by the Community Radio Toolkit. This is a series of articles to help community radio stations analyse their audiences (Gordon, 2010).

Meadows et al's work was examined with a view to developing and evaluating its use in the United Kingdom, enabling "baseline" research of the United Kingdom's community radio sector and audience and also testing the Australian research model in a second environment. The key point of Meadows et al's methodology is that the research was conducted amongst members of the audience and "Key People" from the audience, (Meadows et al, 2007 pp. 22-24) not station staff, volunteers, stakeholders, regulators or other interested bodies. Their previous work (Forde et al, 2002) had used respondents from these groups but their second piece of research was focused on the audience (Meadows et al, 2007).

It should be noted that although there are similarities between the Australian and United Kingdom's community broadcasting sectors, there are also differences. Australia is a large country geographically, putting a high value on local broadcasting as communities may be some distance from the centres of politics, commerce and culture. The power of community radio transmitters is typically much greater, for example the transmitter power of *Radio LaB 97.1FM* is 25 watts while a similar station in Sydney, *Eastside Radio*, transmits at 250 watts. There is also some central funding in Australia for specific programming for indigenous, ethnic and print disabled listeners. However, the most significant difference is that the Australian community radio sector is well established at around 35 years old and the United Kingdom's sector much newer. Many stations in the United Kingdom have only been broadcasting for one, two or three years.

The United Kingdom pilot of audience focus groups based on the Griffith researchers study was conducted using two community stations, *Desi Radio* and *Siren FM*. The stations were approached and asked if they would participate. Both stations were very pleased to do so. They accepted the request that they organise the focus group of listener participants and were aware that no staff or volunteers could take part. They gathered a group of listeners by a combination of on air announcements, requests to listeners who contacted the station for other reasons, and approaching known listeners. It is recognised that a limitation of the methodology is that the focus group participants will tend to come from a group of enthusiastic listeners, who have self-selected to take part. Therefore, the research will not answer the question, "Why aren't you listening?"

The Griffith researchers had asked the focus groups for topics for discussion at the start of the session. However they had found that a number of topics recurred in the sessions. Consequently in the United Kingdom focus groups the Griffith themes were taken as the areas of discussion but with the offer to talk about any other matters that the participants wished to. The Griffith topics were:

The accessibility of station for the listeners:
Can listeners make contact by phone, email, text message, walking in the door?

Presentation and style:
How do listeners regard their relationship with the station broadcasts? Do they feel that the presenters are "friends", knowledgeable, representative of the community?

Local news and information:
Do listeners choose the station for its "local" information?

Music:
What do listeners like or dislike about the music programming? Is it culturally defining rather than simply something that happens between other elements?

Diversity:
Are listeners aware of the communities within communities the station broadcasts to? Do they appreciate the variation of the programming for the different listeners served?
(Meadows et al, 2007, pp. 76-77)

The United Kingdom focus groups were conducted by a moderator and note taker and recorded on an audio recorder. All participants were given a consent form to read and sign and one to keep. This gave the general outline of what the focus group was about, how the information was to be used and expressed a loose code of behaviour for them. The recordings were transcribed for analysis and the participants were anonymised in the transcription. No reward or incentive was offered to attend. The participating listeners who had agreed to take part were welcomed and offered refreshment and generally made comfortable. The entire session was run by the moderator and note taker, who introduced themselves and explained the format of the session. The participants were not invited to introduce themselves as it was felt this better stressed the anonymity of their comments. The consent form also asked that participants respected each other's anonymity. No one connected to the radio station was in the room during the focus group sessions.

The moderator introduced each topic and everyone in the focus group was given the chance to give an opinion. Open discussion was encouraged but the moderator was careful to ensure that vociferous participants did not dominate the discussion. The note taker took notes as to what was said, using abbreviations for the participants, for example M1 (male 1) and F2 (female 2). This was helpful when listening back to the recording in order to transcribe it. They did not attempt to write verbatim, but the key points were noted. The session ended by thanking the listener participants and explaining how valuable their time and contribution was for the station and for all the other listeners.

Following the focus group informal immediate feedback was given to the station managers based on the notes and experience of what happened. Care was taken not to identify individual participants. The recording was then transcribed and this was used to analyse the particular topics and check that issues were not overlooked or missed. The Griffith team had used 'Nvivo', a software analysis program, to analyse the focus groups conducted. Software analysis was not used in the United Kingdom as this was a small scale project. It was also felt that individual community stations would be unlikely to have access to more sophisticated analysis packages, should they wish to follow the same methodology.

Results

The results provided here is an assessment of the methodologies rather than the actual data collected. However data and quotes are used to illustrate and highlight the methods used.

Quantitative results

The quantitative audience survey detailed above has been used annually by students at the University of Bedfordshire (originally known as University of Luton) since 1997. The results have varied between 30 per cent and 65 per cent of the respondents listening during the previous month (Gordon, unpublished material, 1997-2011). Combined with statistical demographic data this suggests a local audience reach of around 10 -12 per cent of the population of Luton.

After publication of the methodology used by these students in 2006 (Gordon, 2006) other stations have used the method or adapted it for their own purposes. *Future Radio* in Norwich has been one of the most diligent community radio stations in conducting organised audience research. They have alternated quantitative non-probability surveys in 2008 and 2010 and qualitative online research in 2007 and 2009 (Ward, 2007, 2008, 2009, 2010).

Ward (2008, 2010) has increased the number of respondents, using 546 in 2008 and 283 in 2010. This increases the confidence in the results, although Ward does point to the difficulties for a station such as *Future Radio* that serves a wide general audience of all age groups. In order to properly survey the audience they need a high number of respondents and also to ensure a gender and age variation approximating the local demographics (Ward, 2008, p. 3).

A particular feature of their research is that Future Radio has actively used their research to change the stations method of operations.

In 2008 the recommendations of the research included:

"The findings suggest the need for more off-air promotion about Future Radio, its purpose and the content and timing of the shows. Staff, volunteers and regular listeners should be actively encouraged to get involved in promoting Future Radio, again explaining the range of programmes on offer, as word of mouth is one of the best sources for gaining new listeners.
Consider adapting the research questions to bring them more in line with those used in the RAJAR. Though any change to methodology will make it harder to compare new listener figures with the previous ones for Future Radio." (Ward 2008 pp. 7-8)

In 2010 the research found that:

"When prompted, 53 per cent of all respondents claimed that they had heard of the station which is an improvement on the previous 2008 survey where 43 per cent stated they had heard of the station. 61 per cent who had lived in the total survey area had heard of the station and younger people were more likely to have heard of the station." (Ward 2010:1)

It was also found that 35 per cent of the respondents had also listened at some point, which represented a significant audience increase compared to the previous research conducted in 2008. The research was used to enhance the station's service to the community and having identified that their potential audience had not heard of them, the station took some trouble to increase the local community's awareness of the station.

Qualitative results

The pilot focus groups were reviewed as to the effectiveness of the methodology rather than the results it produced. The results were examined to discover if the methodology was sound, the topics appropriate, if there were additional areas that need to be covered and whether the participants were able to discuss their pleasures and concerns as listeners.

It was found that the methodology was generally sound. Focus groups are a well established way of unpacking numerical data and giving a richer, personal and anecdotal interpretation of the results (Meadows et al, 2007). The community radio focus groups were well attended; attendees appreciated being asked their views and were willing to participate in the discussions. Respondents fully realised that they represented the audience and took the role seriously. They also valued that the station was asking them their views.

The respondents in the groups were offered the opportunity to extend the discussion and to express views on areas not prompted by the moderator. They did not wish to and expressed the view that the discussion had fairly represented the areas that they had views or feelings about, including some areas that they had not been aware of, for example the notion of the station broadcasting to several adjacent yet distinct communities, or "communities within communities". This was noted in both groups.

On some occasions the moderator needed to encourage participants to provide detailed responses, for example their appreciation of the music played. The participants were surprised that the focus group moderator regarded music as a valuable element of radio, not "just what happens between the speech", and in the case of *Desi Radio* formed a part of the Punjabi cultural identity.

"My mother-in-law likes listening to (the station) more than me. She is almost 94 years old. We listen to it all day long because we know only Punjabi and the programmes are in Punjabi as well. It not only plays songs but also gives interesting information about household things" (Female respondent, Radio Focus Group, *Desi Radio*).

On other occasions it was necessary to explain broadcasting norms to respondents, for example the meaning for the station's output of their "key commitments" (that is, the terms of their licence). This suggests two points; firstly that the moderator needed to ensure that the questions properly briefed participants without using jargon and that self evident broadcasting "truths" are not necessarily obvious to the listener. Secondly, the outcomes of a focus group can suggest where the station needs to inform their listeners as to what the parameters of their programming are. For example if the station has a key commitment to broadcast solely in a particular language, listeners may need to be shown that this is a benefit, rather than a limitation. Focus groups can demonstrate to station staff not only what they are doing well but also where they might develop and also where the audience needs further information to understand what the station is trying to do.

Accessibility was an area where the respondents within the focus groups showed some variation. Individual respondents in both groups indicated that they simply enjoyed listening to the stations and felt no need to get further involved.

> Moderator: "Have you ever tried contacting them [Siren]?"
> Male Respondent: "No I haven't tried contacting them. I like listening to it rather than participating."
> (Male respondent, Radio Focus Group, *Siren FM*)

> "I haven't called any phone-in programmes. I like listening to it passively. I don't like interacting much."
> (Female respondent, Radio Focus Group, *Desi Radio*)

However other respondents stressed the accessibility of their community stations,

> "Siren is involved in a lot of courses and they have lots to do with the community. The aspect of physical accessibility is the most notable feature."
> (Female respondent, Radio Focus Group, *Siren FM*)

> "I have contributed to phone-in programmes. I have called the station a couple of times."
> "I have called the station twice. I haven't contacted them through emails or text messages. In fact I have come to the station once to see the presenter."
> "I have made [music] requests through phone-in programmes."
> (Respondents, Radio Focus Group, *Desi Radio*)

This is notable in that it had been felt that a limitation of this type of research was that the respondents would tend to be enthusiasts. Although this is undoubtedly true, they were not all listeners who felt the need to get in touch with the station actively. Community radio broadcasters may feel that they are not successful unless they have considerable listener participation, but these results would suggest that this is not necessarily the case.

Future Radio have also conducted qualitative research but via an online survey accessed by their website or by an email link. This was done in 2007 and 2009 and they have obtained useful results from this method of collection, whilst acknowledging that it does require participants to have the use of the Internet. In 2007 their survey had 253 respondents and in 2009 it attracted 286. One advantage of these online surveys is that it is easy to see patterns developing between one year and the next. These could indicate changes, for example in the 2009 survey it was found that almost 16 per cent of respondents had been listening less than four months, suggesting a growing listenership. However they may also note a static situation, for example the age of their listeners, had changed very little between 2007 and 2009 (Ward 2007, p.4 & 2009 pp. 4-7).

Conclusion

"We all know that community radio achieves more and is about more than just delivering audiences to a station. Community radio is about delivering new skills, a sense of localness and relevance and a different voice in the often vacuous media landscape."

"…conducting qualitative and quantitative research at your station … isn't *just* about proving audience figures, it's about proving what you *say* you do." (Shember-Critchley, 2010)

Quantitative data, the simple enumeration of listeners gives credibility to a station and demonstrates that it has an audience who have tuned in to it. Qualitative information will then indicate why the audience is listening and what they are gaining from the experience of choosing that community radio station over other available options. It can also suggest what the station might do to better serve its listeners and possibly bring the service to other listeners.

It must be acknowledged that the methodologies described of small scale audience research have limitations in terms of their validity. The quantitative sample is small, and the focus groups are likely to be garnered from committed listeners. However, when the results of these research methods are combined with other means of audience tracking they gain reliability and validity.

As well as formal methods of research there are many other indicators that can enrich a stations understanding of their audience. The United Kingdom's radio regulator compiles an annual report about the community radio stations that have been on air for at least a full financial year. In 2010 it was reported that:

> "Station's research sources include listener correspondence such as emails, social networking, station website forums and phone calls. Some stations use outside broadcasts and local events as opportunities to create a more personal dialogue with their listeners. A small number of stations carry out more formal research projects; on their websites, through street surveys and occasionally with the use of third party companies." (Ofcom 2010, p. 41)

Some of these methods carry more credibility and confidence than others, but are useful ways of triangulating results. For example an unconfirmed adage amongst community radio broadcasters is that a station can assume that it has 100 listeners per phone call received by the station. Common sense would indicate that this is not reliable based on a programme that actively encourages phone participants compared to one that is easy listening music. However in one of the earliest student conducted surveys at the University of Luton temporary station the "audience research indicated an 11 per cent reach of the population and we logged 200 calls over 28 days giving a ratio of 1 call to 99 listeners" (Gordon, 2000, p. 87). The adage may be a helpful indicator over the whole station output and a defined period of time, such as one month.

A further consideration is how the audience is likely to contact the station and making this available to them. A community station for young people, such as *Takeover Radio*, based in Leicester, has a large number of "friends" on Facebook. (http://www.facebook.com/takeoverradio). This method of communication clearly appeals to their young audience. However, a station serving an older and more mature community, such as *Angel FM* in Havant, frequently receives traditional postal correspondence from listeners as well as emails and phone calls (Smith, 2010).

The quantitative method described is very simple to put into place and use several times a year. However, as *Future Radio* has found using around 300 respondents and ensuring that the local demographics are mirrored requires a tighter briefing of the researchers, but gives results with better validity. Done biannually, and alternately with online qualitative research, *Future Radio* is building a very clear picture of their listeners and appreciation of the station's service to them.

The aim of running a focus group amongst listeners is to find out *why* they are listening to the station. This might also give some indication as to why other people do not listen. What is the station doing well and what it might do to enhance its output for its listeners? What a focus group will do is put a "human face" on statistics. It allows for a less structured setting than a one-to-one interview and encourages sharing between participants, which will stimulate recall and memories of programmes that they have enjoyed as listeners to their community radio station.

As with any data collection and handling, community researchers needed to be briefed as to ethical considerations, including both the respondents and the actual data itself. Stations should beware of gaining large amounts of material that the researchers do not know how to handle, what it means or where to go with it after the initial collection. Raw quantitative data needs to be analysed in conjunction with local demographic figures to make it meaningful; a focus group may give flattering statements about a station, but it is not statistical and cannot be quoted as statistically valid.

Community radio globally is frequently a third tier of radio broadcasting after larger public and commercial services. It broadcasts to small audiences, who otherwise might not have a radio service; it covers news and information for remote or distinct locations and it provides an outlet for minority cultures not available elsewhere. But, community radio does not function in a vacuum and inevitably needs to gain resources to support its staff and facilities. In some areas community radio may also need to justify what it does and show its worth or even show that a demand for the type of broadcasting it provides exists. Audience research will help a community radio station demonstrate its effectiveness to organisations that want to know what it does and it may improve the chances of attracting funding from charitable bodies and other non-governmental organisations. Most importantly, audience research will make a clear statement to its listeners that it values and cares about them.

References

Berrigan, F. J. (ed.) (1977) *Access: Some Western Models of Community Media*, UNESCO: Paris.

—. (ed.) (1981) *Community Communications: the Role of Community Media in Development: Reports and Papers on Mass Communications, No. 90*, UNESCO: Paris.

Community Broadcasting Association of Australia. (2009). *Community Broadcasting Database: Survey of the community radio sector; 2007-2008 financial period; public release report*, CBOnline: Sydney. Retrieved August 10, 2011, from http://www.cbonline.org.au/index.cfm?pageId=37,0,1,0

Community Radio Order (2004). Retrieved April 11, 2011, from http://www.legislation.gov.uk/uksi/2004/1944/body/made

Everitt, A. (2003a.) *New Voices: An Evaluation of 15 Access Radio Projects*. The Radio Authority: London.

—. (2003b.) *New Voices: An update*, Radio Authority: London.

Ewart J., Forde, S., Foxwell, K. & Meadows, M. (2007). *Community Media Matters: An Audience Study of the Australian Community Broadcasting Sector*. Brisbane: Griffith University.

Forde, S., Foxwell, K., & Meadows, M. (2002). *Culture, Commitment Community: The Australian Community Radio Sector*. Griffith University: Brisbane.

Foth, M., Hearn, G., Lennie, J. & Tacchi, J. (2009). *Action Research and New Media*, Hampton Press: New York.

Girard, B. (1992). *A Passion For Radio: Radio Waves and Community*, Red Rose Books: Montreal.

Gordon, J. (2000). *The RSL Ultra Local Radio*, ULP: United Kingdom

—. (2006). Is anybody out there? Audience research for community stations. *Airflash*. Summer 2006, CMA, United Kingdom. Also available at http://www.communityradiotoolkit.net/. pp. 20-21.

Gordon, Janey (2010). How to run a listener focus group - Advice for community radio stations. In Shember-Critchley, E. (ed.). (2010). *Prove It: Community Radio Toolkit*. Retrieved April 11, 2011, from http://www.communityradiotoolkit.net/news/welcome-to-prove-it/

Keith, M. C. (1995). *Signals in the Air: Native broadcasting in America*, Praeger: Santa Barbara.

Lewis, P. (ed.) (1993). *Alternative Media: Linking Global and Local Reports and Papers on Mass Communications, No. 107*, UNESCO: Paris.

Lewis P. & Booth J. (1989). *The Invisible Medium: Radio*. Macmillan, London.

Lewis, P., Slater, D., & Tacchi, J. (2003). Evaluating community based media initiatives: an ethnographic action research approach. Retrieved April 11, 2011, from http://portal.unesco.org/ci/en/ev.php-URL_ID=15731&URL_DO=DO_TOPIC&URL_SECTION=201.html

Ofcom (2010) *Annual Report on the Sector 2009/2010*, Ofcom, United Kingdom. Retrieved April 11, 2011, from http://stakeholders.ofcom.org.uk/broadcasting/radio/community/annual -reports/09-10/

Office for National Statistics (2010). *Mid Year Population Estimates 2009*. Retrieved April 11, 2011, from http://www.statistics.gov.uk/statbase/Product.asp?vlnk=15106

Putnam, R. D. (2000). *Bowling Alone: The Collapse and Revival of American Community,* Simon and Schuster: New York.

Research Works Limited. (May 2004). *Ofcom, Community Radio Final Report*, Ofcom, United Kingdom.

Shember-Critchley, E. (ed.) (2010). *Prove It: Community Radio Toolkit.* Retrieved April 11, 2011, from http://www.communityradiotoolkit.net/news/welcome-to-prove-it/

Smith, C. (2010). *Angel Radio Annual Report*. Retrieved April 5, 2011, from http://angelradio.co.uk

Takeover Radio 103.2, Facebook home page. Retrieved April 5, 2011, from http://www.facebook.com/takeoverradio

Ward, E. (2007). *Future Radio Listener Survey Report 2007*. Blueprint Research: United Kingdom.

—. (2008). *Future Radio Listener Figures Report Summer 2008*. Blueprint Research: United Kingdom.

—. (2009). *Future Radio Listener Survey Report 2009*. Blueprint Research: United Kingdom.

—. (2010). *Future Radio Listener Figures Report April 2010*. Blueprint Research: United Kingdom.

CHAPTER FOURTEEN

THE NET-AMORPHOSIS OF RADIO
AS A SURVIVAL STRATEGY

PIERRE C. BÉLANGER

Being the largest radio broadcaster in your country is one thing. Being the largest and aspiring to be the most popular in the eyes of the public and the reference for your industry peers is a challenge that requires vision, talent, discipline, creativity and flawless execution. It also requires an acute sense of what emerging distribution platforms can do to extend your brand presence into the fledgling digital venues that traditional media are learning to colonise.

This chapter provides a brief overview of radio in Canada, before describing the rationale and strategies by which Canada's largest private radio broadcaster, Astral Radio, is transforming its business model to adapt to the imperatives of the digital environment and the inevitable set of codes, practices and listening behaviours that entails. Two recent initiatives will serve to illustrate the extent to which Astral Radio has decided to go "not where the puck is", as every Canadian hockey player knows, but rather "where the puck is going to be".

As such, a five-year CAD$6 million agreement was signed in the winter of 2010 with Emmis Interactive, a fully-owned subsidiary of Chicago-based Emmis Broadcasting. The agreement is designed to significantly reposition Astral Radio's image on the web through combination of a new look, dynamic content and functionality, more integrated interplay between the on-air hosts and programming elements, and innovative advertising concepts. This undertaking constitutes not only a major investment in operations, but in organisational culture and human resources as well; 55 new positions are created across Astral's radio properties throughout Canada, bringing the total tally of the venture to CAD$25 million over the five-year covenant.

In parallel, Astral Radio created the Digital Platforms Lab, which serves to assess the prevalence, significance and likely repercussions of emerging digital devices and applications on the future of radio as we used to know it. The Lab's activities are conducted with the collaboration of some 15 "digital natives" aged between 20 and 30 years old, considered to be innovative users in digital parlance. The Lab's *raison d'être* rests upon a desire to better understand the specificities associated with the digital distribution platforms in order to adapt and reinvigorate radio's pertinence in the media landscape of the "Gen-Yers".

The Canadian radio market

Canada is known for its geographical expanse. Its 33 million inhabitants are scattered over a territory of 10 million km^2 bridging the Atlantic, Pacific and Arctic oceans, with 75 per cent of the population living within 150km of the United States border. With an average of only 3.5 inhabitants per square kilometre, Canada has one of the lowest population densities in the world.

Canada's 1,221 licensed radio and audio services make it one of the most extensive community-based broadcasting systems among the G20 countries. From a language standpoint, 75 per cent of the radio and audio services are broadcast in English (n=910), 22 per cent in French (n=265) and 3 per cent in another language (n=46) (CRTC, 2010). With the exception of the United States, with more than 14,000 licensed stations, Canada has more radio stations than France (886), Australia (688), the U.K. (510), Japan (368), Germany (278) and Italy (202) (ACMA, 2009; CRTC, 2009). Of these Canadian services, 98 per cent are over-the-air and 60 per cent are provided by private commercial broadcasters, while the national broadcaster (CBC/Société Radio-Canada) accounts for 9 per cent. The remaining 31 per cent consist of religious, community, campus, aboriginal and other radio and audio services.

One of the defining characteristics of the Canadian radio market is the fact that more than two-thirds of the private stations operate in "small markets", i.e. with a population base of less than 250,000 people. The remaining third are located in metropolitan areas of 500,000 people and higher. From an economic standpoint, the 644 private commercial radio undertakings that were operating in 2009 generated some CAD$1.5b, a 5 per cent slowdown over the previous year.

One cannot speak of the Canadian radio market as a single commercial entity, as the linguistic idiosyncrasies of the English and French-speaking markets [1] not only bring about significant business differences but, regulatory ones as well. The English-speaking market is dominated by five operators which together account for 68 per cent of revenues. They are: Astral Inc with 21 per cent of English-language market revenues; Corus Entertainment Inc at 16 per cent; Rogers Broadcasting Ltd with 14 per cent; CTVglobemedia at 11 per cent and Newcap at 6 per cent. The French-language private radio market is concentrated around three operators which together account for 81 per cent of total revenues. Astral Inc is the dominant player with almost half the revenues (45 per cent), followed by Corus Entertainment Inc. at 21 per cent and Cogeco Inc at 15 per cent (CRTC, 2010).

Not surprisingly, average weekly hours dedicated to radio listening in Canada were down across all age groups by a collective 3.2 per cent in 2009 to 17.7 hours. More troubling are the comparative numbers for 2008 and 2009 for the following three demographic groups: Teens 12-17 show a year-over-year decrease of 6.2 per cent at 6.8 hours a week; Adults 18-24 decreased by 8 per cent at 12 hours a week; and Adults 25-35 decreased by 4.1 per cent with 16.6 hours a week going to traditional radio listening (CRTC, 2010). Such numbers only serve to reinforce what everyone in the industry has feared for many years – the slow but gradual erosion of the younger public to the benefit of digital distribution platforms.

The 2009 Canadian tuning trends continue to favour the private commercial broadcasters, which together captured 79 per cent of total tuning. The five largest operators listed above maintained their 59 per cent share in the English-language markets while the leading three French-language radio operators are keeping pace with a 60 per cent share in theirs.

Broadcasting next to the powerful American entertainment industry led the Canadian government to impose a series of quotas with regard to the proportion of local artists aired on Canadian airwaves. As such, the regulations require that at least 35 per cent of popular musical selections played during each broadcast week on English-language stations be Canadian. In addition, to prevent Canadian selections from being played

---

[1] Based on the 2006 federal census, Canadians who declared French as their mother tongue represent 22.1 per cent of the total population (Statistics Canada, 2006).

outside of the prime time periods, the law requires that the 35 per cent quota be aired between 6 a.m. and 6 p.m., Monday through Friday during any broadcast week.

In addition, because the government considers that French-language private stations have a pivotal role to play both in promoting and enriching the cultural needs and interests of their audiences, regulations require that at least 65 per cent of popular vocal music selections aired during each broadcast week be in French. Even more stringent than the quotas imposed on the English-language stations, the 6 a.m. to 6 p.m. requirement is set at 55 per cent of French-language selections. If taken literally, however, this requirement could result in a radio station playing only artists from France, Belgium, or another French-speaking country, since the French Canadian market could hardly meet the demand without an annoying degree of redundancy. In order to prevent the exclusion of French-language vocal music produced in Canada, the law further specifies that no less than 35 per cent of popular music selections aired weekly by French-language commercial stations be Canadian. For all intents and purposes, the requirements imposed on Canada's French-language stations constitute by far the highest linguistic or national music quota in the world (CAB, 2008).

Astral Radio

By acquiring the country's largest English-language radio broadcaster in 2007, Standard Broadcasting Corporation, Astral Radio became Canada's *de facto* leader in both official languages with its stable of 62 English-language and 21 French-language stations scattered over eight provinces with a concentration in the provinces of British Columbia, Ontario and Quebec.

Astral Radio owns some of the country's best known brands: Virgin, NRJ, Rock Détente, EZ Rock and The Bear. Taken together, Astral's 83 radio stations reach some 17 million Canadian listeners every week in 50 markets with a 15 per cent share of tuning to English-language radio and a 28 per cent share to French-language radio stations. These numbers make Astral the leader among private commercial radio operators in Canada.

Two recent developments have contributed significantly to the rise of Astral Radio's profile, both nationally and internationally. On August 25, 2008, Toronto's MIX 99.9 became the first North American commercial

FM radio station to be rebranded as Virgin Radio. In January 2009, three more stations in Vancouver, Calgary and Montreal adopted the Virgin on-air signage with Edmonton becoming the fifth to join the Virgin format in February 2011. Why the conversion?

The key factor was the desire to be the first radio group in America to partner with the notoriety and appeal of a brand as renowned as Virgin. As one of the fastest growing companies in Sir Richard Branson's Virgin group, Virgin Radio International has been expanding in recent years into Italy, France, Dubai and India. Astral is convinced that the Virgin brand will resonate strongly with its English-language markets, with the rebranded stations having the potential to reach new listeners and thus positively impact advertising revenues. Moreover, the association offers a host of opportunities to leverage the global and local power of the Virgin brand.

Encouraged by the Virgin experience and seeking to freshen its eight year old Radio Énergie brand that was showing signs of fatigue, Astral once again reached out to a European partner in August 2009 and renamed its 10 Radio Énergie stations in the province of Québec after France's largest radio network NRJ, a natural association considering the proximity of the two brands' phonetics. Along with the emblematic black panther, the new NRJ Canadian stations have adopted their partner's jingles and slogan while maintaining their autonomy with regard to playlists, programming choices and hosts. By joining the NRJ family, Astral becomes part of a truly international network of stations. With its 280 stations spread over 16 countries that include Sweden, Belgium, Germany, Russia, Ukraine and others, along with its television properties, concert production and recording units, NRJ offers Astral a wealth of content that other Canadian broadcasters can only dream of. As such, when U2 or Madonna gives an interview to any one of the NRJ European sister stations, Astral has access to it. Likewise, any material that Astral owns about French-speaking Quebec artists is made available to interested NRJ partners, thus creating a considerable window for the promotion of Canadian talent abroad.

Redesigning Astral's digital persona

By acquiring Standard Broadcasting Corporation's radio assets, Astral Radio was inheriting its collection of antiquated content management systems (CMS) that only added complexity to what Astral was already using. Dealing with a variety of CMS within the same company is a costly, time consuming, resource heavy infrastructure that considerably hinders a company's ability to innovate.

Soon after the hiring of a chief digital strategies officer in the summer of 2009, Astral elected to streamline its web activities and migrate its three main CMS and related sub-systems into a common web CMS, known as an engagement management system. That decision was motivated both by the urgency to bolster many of the group's then stale-looking web properties, and by the need to revitalise the entire collection of websites so that they could be a contributing factor in the digital sales strategy that was being developed. The objective was clear: to build a nimble business model capable of expanding, responding and adapting to the changing online landscape while generating revenue through multiple sources. By proceeding with a massive overhaul of its web operations, Astral was counting on making significant gains on the way its business partners would now perceive its capacity to blend on air and online strategies.

The decision to proceed with a company-wide integrated web CMS quickly became a necessity. Everyone in management saw the tangible benefits that would ensue, and acknowledged that fast tracking the project was the only way to close the gap with the competition and send a clear signal to the industry that Astral was making a commitment to extend its leadership in traditional radio over to the digital platforms arena. However, what was considerably more difficult to agree upon was whether to "build" or "buy", i.e. build the system internally or outsource it.

Astral's internal interactive unit [2] was adamant about its ability to deliver a full-fledged, custom-designed proprietary system that would be the envy of any radio broadcaster. It had the wherewithal, it claimed, to prove itself and re-establish a reputation sullied by many years of unmet promises and what it considered to be unreasonable expectations considering the paucity of resources available. A costly failed venture into a subscription streaming music service four years prior had nurtured a profound scepticism about the launch of any further home-grown initiative. The overriding management posture is best summarised as "Been there, done that, and... still paying for it".

---

[2] Information contained in this section was obtained through interviews with two senior members of the Astral Radio interactive team.

In such a low confidence internal environment, combined with the pressure to get to market as soon as possible, the "buy" option easily won over. In February 2010, Emmis Interactive, a subsidiary of Emmis Broadcasting from Chicago, Illinois, was selected as the sole vendor of the web CMS. Emmis committed to fully integrate Astral's various CMS models into its BaseStation system, guaranteed its fluid functionality and, most importantly, promised the launch of a redesigned version of the radio group's 83 websites by the end of the summer – no small feat considering the number of radio brands involved and the French and English language issues. Internally, the project became known as code-name FUSION: Force, User, Synergy, Interactive and ON, as in "all engines running".

Emmis' forte as a service provider resides in the fact that it is, itself, an off-shoot of a radio operator. Hence, Emmis understands the challenges facing Astral, its needs and the highly competitive nature of the environment it lives in. And because the vendor has been there before, it knows that by definition, Astral's core competency is primarily in managing its stations, increasing sales, and developing new concepts, new promotions, new talent, etc. In today's context, nobody disputes that taking on digital responsibilities is the natural thing to do. But doing so comes at a price – sometimes steep, generally frustrating, and always much too time-consuming. The learning curve is long and tedious; a product's iterations seemingly never ending. Good and reliable staff are costly, have a different sense of loyalty and tend to get poached by competitors. Suddenly outsourcing becomes appealing, especially if it comes with an enticing package of services.

The chosen external provider had a very good story to tell. For starters, the platform it had developed for Chicago Q101 was being used in some 20 American markets where the proportion of web-driven advertising revenues to total station sales averaged 9 per cent, even climbing as high as 16 per cent in one case; a vigorous number that would get any radio operator's attention. What made those numbers even more enticing was the way by which they were obtained: they were the direct by-product of a carefully planned business model and developmental strategy that made the BaseStation proposition all the more compelling.

At the basis of the Emmis model, and one of its fundamental tenets, is the notion of local station empowerment which runs in direct opposition to syndicating content or centralising web services. While Astral's stations used to rely on a team of English-language web content producers, designers and programmers based in Toronto with French-language stations on staff located in Montreal, Astral bought into a market-based paradigm where each station is responsible for the daily operations of its website and, ultimately, for its success or failure. In essence, the radio content management model is now being extended to the websites.

It is now incumbent upon each local market to generate and publish fresh targeted local content on a daily basis on Astral's interactive platforms. To that end, two new positions were created and a third one redefined: the digital content producer, the digital account manager, and the brand director.

The digital content producer's (DCP) primary role is to make sure that all websites under their care spawn locally pertinent and frequently updated material. The DCP works closely with the digital account manager to produce and implement custom advertising solutions capable of delivering strong return on investment for the buyer. He collaborates with programming and promotions departments to build and maintain a meaningful and viral database of the stations' members. The DCP is also responsible for the training and support of all on-air staff. As such, the DCP must ensure that during their workday, personalities on Astral Radio stations not only create enticing content but actively promote it on-air. Content might come in a variety of forms including blogs, and moderating chat modules, driving conversation and engagement in social media platforms which are now perceived as valuable brand extension and marketing platforms. Regardless of the form it is conveyed in, content must ultimately be designed to establish involvement and collaboration with listeners and foster listening appointments.

The digital account manager (DAM) is the new in-house expert at delivering compelling pre-sell presentations and is accountable for ensuring that the right message gets delivered to the right prospective client. As such, the DAM works closely with the local DCP in overseeing successful launch and execution of campaigns as well as post-campaign reporting. In addition, the DAM is responsible for the on-going digital coaching and training of account executives.

The brand director (BD) of a local market is the new title given to the programme director and is accountable for Astral's assets across all platforms. As the steward of his brand(s), the BD has final say over on-line content as well as any programming, promotional or sales partners associated with any of the brands present in a given market. The BD is responsible for content, layout, and promotion, both on-air and online, and must ensure the best online experience for his market stations' fans. Ultimately, it is the BD who is responsible for achieving targets for page views and unique visitors, while sharing web revenue responsibilities with the sales director.

As part of the local responsibilities now imparted to each radio market, all stations are required to air a minimum of four web promos daily and provide listeners with two to four reasons why they should visit the station's website. In addition, every station is expected to devote eight seconds within the body of the promo to support a national sales promotion.

For Astral Radio, partnering with Emmis Interactive was much more than simply buying a technological platform that would make its 83 websites look, feel and function better. Astral Radio was getting a solution; a suite of services that promised to invigorate its web properties and stimulate their full revenue potential. Emmis' model is predicated upon a three-pronged strategy: 1) provide your local markets with qualified web and sales staff, 2) equip them with the right toolset, and 3) allow for sufficient on-site training to take place. This is, on the face of it, a simple approach that nonetheless calls for a massive financial commitment in light of the number of new hires required, the redesign of the website templates and functionalities and the associated coaching costs. After the usual back and forth on the specifics of the contract, the agreement was finally sealed in the spring of 2010. Astral Radio was now committed more than ever before to building a two way thoroughfare between its on-air and online properties.

Operationalising the solution

Senior management was adamant: now that the agreement is signed, all 83 sites must be revamped, and web and sales-dedicated staff must be hired and trained before the start of the fall season. To coordinate the mission at Astral, a triumvirate was formed, comprised of: a) a project leader responsible for the financial and organisational elements of the project; b) a former station director, highly regarded and perceived as "one of them" in both French and English markets, in charge of evangelising throughout the various markets, and c) a digital platform officer overseeing the technological components of the new solution being implemented.

And so the "road show" began. Starting in March, every radio market was first visited by two members of the triumvirate whose objective was not only to explain the details of the upcoming web overhaul but, more importantly, to tell them about the ways in which they would be supported in making the transition. Because Astral Radio was entering uncharted territory, it was felt that in order for this project to develop any traction, everyone needed to understand the nature of the changes coming, the extent of their involvement, as well as management expectations.

Attending those meetings were the market stations' directors, the programme directors and the sales directors. They were informed that starting now, head office was freeing up the required budget so that each market could hire a webmaster, a specialist in web sales – also known as the digital account manager – and could convert the programme director's job description into that of a brand director. Because Astral was buying the rights to use Emmis' platform, business model and philosophy, the latter was directly involved in the hiring process, advising Astral's human resources and interactive media representatives on the types of candidates best suited to implement BaseStation. In most markets, webmasters and brand directors were provided with one day of on-site technical training where guidelines were spelled out and design and framework standards demonstrated. French-language markets were allocated one and a half days in order to account for *in situ* translation services. Sales teams were run through a full day of workshops. At the end of the road shows in June, a grand total of 1,000 hours of training had been dispensed. This clearly was one of the key "on the ground" features provided by Emmis that swayed Astral to come on board.

Another was the frames that Emmis was developing for Astral Radio's various music formats such as rock, news and talk, adult contemporary and classic rock, in order to facilitate the job of the webmasters. As such, the latter find themselves working more as layout designers than as web programmers per se, since they can dedicate the better portion of their time working on adding content elements to the site rather than on programming codes. The frames allow for a standardised size and location for adverts throughout the company's web properties, regardless of the music format. Likewise, many of the visuals developed for various concert tours, music releases, contests and promotions can be shared by all, significantly increasing the efficiency and coherence of the collective effort.

As with any operation of this magnitude, with the kinds of repercussions that it unavoidably provokes on the organisational structure of the company, such a venture is launched because it first and foremost makes financial sense for the buyers. Considering that before the launch of BaseStation, Astral was generating less than 2 per cent of its total advertising revenues from its web properties, it is estimated that should the new strategy succeed in raising the proportion of web-based revenues to 5 per cent, the investment will be recouped within three years and become accretive on EBITDA[3] in the second year.

Although it is premature at this point to determine whether the project has proven to be a success, early indicators are sending all the right signals. In an internal memo sent to employees on December 7, 2010, Astral announced that sales numbers for the first trimester under the FUSION strategy had come in at 112 per cent of set target, an achievement that, although too premature to make any claims, is the sort of signal everyone was hoping for. Six months into the rollout, sales growth is coming in at 78 per cent year-over-year, another key indication that the execution is on the right trajectory. For year one, the corporate objective is to have web-based activities generating an overall average of 2.5 per cent of broadcast revenues, with a range set at between 2 per cent and 5 per cent. Without getting into the specifics, the target for 2011 is to increase web-based sales by 75 per cent over 2010, by 300 per cent by 2012 and by approximately 450 per cent in 2013. As aggressive as these targets might seem, almost every station gave its unconditional support. It speaks to the natural entrepreneurial DNA of private commercial radio operators who feel that, given the set of resources that they are now provided with, these sorts of numbers are realistic and achievable, barring any major economic slowdown like the one the western world is still painfully working its way out of. Individual radio markets are now the masters of their own destiny. The targets are clear. The strategy has been demonstrated to work efficiently in over 20 US markets. Corporate support is tangible and unwavering as illustrated by the way sales bonuses have been restructured.

Anybody who works in the broadcast industry knows only too well that the much heralded migration of advertising money toward new digital platforms is more than just a passing fad. In its most recent report on the advertising industry, eMarketer (Online Ad Spend, 2010) forecasts a 10.5 per cent increase in US online ad spending in 2011, followed by double-digit growth every year through 2014. David Hallerman, principal analyst

---

[3] EBITDA: *earnings* before *interest*, *taxes*, *depreciation*, and *amortisation*.

at eMarketer, sees this trend as a reflection of "how most forms of Internet advertising are now seen as more of a 'sure thing' than most traditional media". In this context, it is not surprising to see Astral getting in on the act and significantly revising its sales strategy.

When adding to those online advertising projections the fact that the mobile screens of smartphones and tablet media are developing into credible advertising venues, you have a set of conditions that warrant a new compensation model designed to hearten your sales force to promote the on-air/online combination. Under the terms of the deployment of the FUSION project, any web sale generates an extra 2 per cent commission for the sales rep and the director of sales. The station manager and the brand director also benefit, although their compensation is directly linked to the station's EBITDA. Contrary to many other incentive programmes, people involved in web sales get a bonus when they reach their market web targets even if their radio numbers have not been attained. Under this web-minded logic, the bonus gets more lucrative when both the pooled web and radio targets are met. This sort of gratification-by-premium programme is the time-tested formula to boost sales as reps and their supporting cast are naturally conditioned to respond to this sort of stimulation.

Lessons learned

Ostensibly, it is still too early in this major digital overhaul to draw any conclusions. But as the project unfolds, as people get comfortable with their role and expected contribution, as web traffic and sales numbers move up and technical glitches get ironed out, is it possible to take a few steps back and look at the road travelled over the last year with any objective distance?

In hindsight, although everybody recognised that something had to be done to steer the ship in a more progressive direction, senior management needed to see the evidence that FUSION was the way to go. Factors fuelling the scepticism were: the business model seemed overly optimistic; the 100-day timeframe allotted to significantly alter the culture of the enterprise appeared short considering the extent and complexity of recruitment associated with the project; the number of people whose skills were deemed to be no longer compatible with the FUSION requirements; the level of pressure to be put on the sales force to make the project work; the capability of Emmis to dispense the required training to the large a number of markets within such a tight schedule; and the design, development and testing of the website templates. These were all valid concerns that required factual and measurable answers.

At the end of the day, implementing a dynamic web strategy to freshen your online image, and combining that change with the imperative to increase advertising revenues because they are facing headwinds in your traditional business units, boils down to being much more a human resource challenge than a technological one. Whichever way you analyse it, the technology behind FUSION remains just another way to support standard web applications, content and services. There is, to be brutally frank, nothing exceptional about FUSION. In that sense, it can be fairly easily mastered and operated. On the other hand, the material posted on the radio websites, its selection, originality, timeliness, effectiveness and attractiveness are all intimately connected to the quality of the people chosen to curate the sites. Without a doubt, the biggest challenge of this undertaking has to do with giving people a chance to prove themselves and contribute. That applies to the brand directors, of course, but equally to the sales reps and the station's management as well on-air talent who all must now learn to do old things in different ways in order to fully capitalise on the radio/web synergy. To remain relevant, radio must fully engage in the battle of mindshare; a battle that is simultaneously being fought at the juncture of traditional, digital and mobile fields.

The capital importance that continuous training and coaching play in the success of such a project cannot be stressed enough. Figure 1 below provides an example of the kind of material that is provided to on-air talent. In this specific instance, it serves as an abridged tutorial to guide and inspire hosts so that social media tools become an integral part of what they do before, during, after and between shows.

Figure 1. Executing social media

F U S I O N: Force / User / Synergy / Interactive

PRE SHOW	YOUR SHOW	POST SHOW	24/24
1. What do we talk about (what's the talk of the town?)	1. I give exclusive information related to my show.	1. I thank the ones that followed.	1. I am listening. I am present.
2. How can I influence the conversation?	2. I bring the listeners/ fans to the station's site.	2. I answer unanswered questions.	2. I start a conversation, I share content.
3. I start the conversation.	3. I have a dialogue with the listeners – I optimize the perception that they influence my show.	3. I continue the conversation.	3. I am authentic, personal.
			4. I increase the number of fans on my personal pages.
Objective	Objective	Objective	Objective
I draw attention.	I connect. I give non-listeners reasons to listen, and listeners reasons to listen longer.	I complete what is unfinished. I give content added value.	I built my personality (my brand) – I increase my credibility, my loyalty (my number of fans).

It is but one of the many aids that Astral regularly produces to acculturate its workforce to the full potential of the web. In the process, this type of material reinforces the notion that the media landscape is changing quickly and that rapid, tangible and decisive diversification remains of the best adaptive tools. Astral complements those instructional guides with a regular internal evaluation that is conducted on the performance of the websites. One such example took place between the months of August and October where seven stations' websites were submitted to a panel of 15 participants aged 20 to 34. In many cases, the new version of the sites selected for this examination had only been running for a few weeks. Nevertheless, management wanted to get a sense of how representatives of the target audience responded to the redesigned version of the sites and the sort of quick fixes that could be made to improve the connection with the users.

Appraising the user's experience

It is always a delicate balance when one has to decide between holding off to launch a new website until it is thoroughly tested and all the bells and whistles are perfectly adjusted, and proceeding with a so-called "soft launch" where the site's functionalities and content categories are not yet finalised but are considered to be at an advanced enough stage to justify hitting the market. Eager to show its employees that FUSION was on schedule and that the newfangled synergy between on-air content and online presence was indeed operational, Astral Radio elected to hit the ground running and go with a discrete, progressive launch and make the required revisions along the way. In doing so, every station could thus boast on air about the new look and appealing user-focused complements to on-air programmes, offer the dominant social media services of the day and, most importantly from a marketing point of view, be available for the start of the new radio season.

Predictably, launching 83 websites in less than three months does not come without a few glitches. Even if many of the sites share a common template by virtue of the fact that they belong to a similar programming format, they nonetheless feature content, contests and local information that remain very specific to each market. Astral knew that it was taking some calculated risks by putting the sites online as soon as they left the production line. But it needed the exposure, felt it had to close the gap with its competitors and wanted to send an unequivocal signal throughout the market that Astral Radio was determined to play a leading and influential role in the transformation of radio.

Astral turned to its Digital Platforms Lab to put seven of those new websites to the test. On average, evaluators were given a 10-day period to explore each site. At the outset, it was agreed that participants would dedicate a minimum of two hours over that period to get acquainted with the sites' various sections and features and thus adequately test their relevance, effectiveness and ease of use. Evaluation questionnaires were filled out upon completion of each 10 day examination period. Two focus groups sessions were held; a first session after the first three sites had been reviewed and a second session after the last four. The sessions were transcribed, and salient observations divided into four action-based categories:

1. I like
2. Elements of the site that should be improved
3. Elements/functions that the site should offer
4. Please fix.

In all, some 273 recommendations were made to optimise the experience of the sites. The essence of those comments was subsequently synthesised into 10 clusters in the report that was delivered to management.

In all likelihood, the majority of the observations collected and which made up the clusters of actionable recommendations listed below would apply to most, if not all, broadcasting websites. In this sense, they constitute an edifying compendium of the impressions that traditional media websites' audio and visual components, content categories, functionalities, and use of social networks and advertising elicit in their users.

1. Quality control

If there is one dominant lesson to take away from the redesign of Astral Radio's websites, it is the importance of setting up a quality control mechanism; a role with responsibility for constantly monitoring the user experience aspects of a station's web presence. Spotting and eradicating glitches must be integrated into the daily routine.

2. Less is more

Evaluators were adamant on this point. They want less scrolling, less clicking, less waiting for pages to load, less repetitive content, and less advertising that is not related to music or programming content.

3. Search tool

The absence of a search tool is a direct by-product of having chosen the "launch-right-away-fix-on-the-run" approach. It was unanimously criticised by the participants who felt they were wasting valuable time locating content.

4. Catch-casting player

Along with an impressive number of archived podcasts, Virgin Radio-France offers a vivid example of an effective tool to quickly jump back and forth between live programming and that which has been aired in the last 24 hours. When showed at a focus group session, this catch-casting player impressed everyone and promptly became one of the most pressing desiderata expressed by the participants. However, depending on the jurisdiction governing copyrights, many radio broadcasters might not have the liberty to use this type of popular feature.

5. Personal information

Most participants resented having to provide personal information and sign up to join a station's VIP club or lounge; they ostensibly fail to perceive any value in the activity. It is doubtful anyone would have registered had it not been for the fact they were analysing the sites, or that they had chosen to enter a contest or elected to join in order to receive exclusive content. The purpose of encouraging audiences to sign up, as anyone in the industry knows, is to build a bank of socio-demographic data in order to focus on specific market segments sought by advertisers. But why fill out a form to request a song when one can type five words on a Twitter account and click "send"? All participants expressed their interest in becoming a fan via Facebook or a Twitter follower; two ways to become part of the Astral Radio family deemed much more efficient.

6. Favicon

Each site's URL should be preceded by a "favicon" for quick visual identification in the users' bookmark lists. This is a simple operation to perform, yet its impact on the perception of a station's website is far from negligible.

7. Sharing

Emblematic of the social media culture, sharing has become the undisputable new black of the digital world. As such, every site element should be made "sharable" via as many of the leading services as possible. One way to achieve this is to make the highly recognisable bookmarking and sharing icon ubiquitous throughout the sites... and watch the content travel through users' personal networks.

8. Twitter

Although it is now customary for a website to display the Twitter icon, in many cases its use is far from optimised. Radio operators need to realise that social networks demand a fair level of human resource to manage, update, respond to and just generally keep the conversation alive. The first step to undertake is increase the number and range of people a station follows on Twitter. Second, internally cultivate a Twitter culture and encourage programme hosts to be active users themselves. This strikes at the heart of social media.

9. Interaction

Because it is what they expect and willingly engage in, users of radio websites want to be able to interact with hosts, programme elements and fellow listeners/users during shows. Allow listeners to share, vote, "like", "bomb"/"rock" or "dislike" any content they find on your website. A series of tactics should be deployed to satisfy this propensity. Location-based social media initiatives should be experimented with as they are gaining traction in the marketing and advertising fields.

10. Feedback Tool

A simple and efficient way to detect glitches and solicit feedback is to post a feedback button on a site and encourage listeners to share their online experience. Users can thus help a radio station's website quickly fix both technical and content-related anomalies, as well as suggest ideas to optimise the general performance and usefulness of the sites.

Over and above the specific areas of improvement that the list above enumerates, it also highlights the desirability of having an on-going formative evaluation protocol integrated into the developmental phases of any website functionality or content element. Considering the importance of the financial investments associated with the development and constant upgrade of a radio station's website and the related contingent expenditures, a permanent arrangement should be put in place to appraise, from a user's perspective, both the level of satisfaction and efficacy of the products' components.

Many major corporations such as Proctor & Gamble, Kodak and Dell among others, have recently created the position of "chief listening officers" (Slutsky, 2010) whose role consists of keeping their ear close to the ground in order to promptly advise on how the company should respond to what clients say about and do with the products that they have purchased. This is the epitome of driving innovation through customer feedback, a trend all the more crucial at a time when social media are amplifying the reach and impact of customer opinion on the products being consumed. In the permanently evolving digital landscape in which traditional radio stations have no choice but to carve a niche for themselves, the relevance of being attuned and responsive to what is being said about you is all the more acute. This is how product and strategic direction should be shaped.

Beyond radio's website presence

Other than the ubiquitous media tablets, few developments are currently getting as much attention as HTML5, the newest iteration of the HTML standard for structuring and presenting content on the web and particularly, what it can do for mobile browsers. Because it is getting late to show up at the apps party, some have started to raise the spectre of a mobile "app-pocalypse" (Leggatt, 2010a), a neologism provocatively constructed by the proponents of mobile web browsers who contend that mobile apps are transitory and only came about because they are the best way yet to circumvent the limitations of smartphones in terms of network bandwidth and processing power.

HTML5 is expected to deliver a much more satisfying experience than apps are currently capable of, given that mobile versions of sites can fluidly work across all platforms and devices, and are not faced with the same longevity issues. With the next generation of faster, more powerful smartphones and the deployment of more robust 4G networks, HTML5 is likely to significantly improve the mobile web browsing experience. If radio stations need evidence of a momentum shift in the area of mobile developments, they have to look no further than dotMobi's study which reports an impressive 2000 per cent growth in the number of mobile-ready websites in the last two years, with 40 per cent of the world's top websites offering a mobile version (Leggatt, 2010b).

Likewise, a recent mobile consumer study conducted by Adobe Systems Incorporated (2010) reveals that when it comes to the mobile user experience, respondents generally favour mobile browser experiences over downloadable mobile app experiences across all four key consumer categories including, media and entertainment, where the ratio is 2:1 in favour of mobile web access.

Although one would be hard pressed to support the apps naysayers in light of the convincingly strong adoption numbers that apps generate, radio operators can no longer dismiss the strength of the mobile browser trend. As such, the industry as a whole would be well advised to catch this rising tide and dedicate the level of resources that it requires. If there was ever a development likely to bring radio fully into the digital and social media age, mobile web browsing is it.

References

ACMA. (2009). *Communications report 2008-09.* Melbourne, Australia: Australian Communications and Media Authority. Retrieved December 2, 2010, from http://www.acma.gov.au/webwr/_assets/main /lib311252/08-09_comms_report.pdf

Adobe Systems Incorporated. (13 October 2010). *Adobe mobile study reveals consumer preferences for accessing consumer products and shopping, financial services, media & entertainment and travel from mobile devices.* Retrieved November 7, 2010, from http://www.adobe.com/aboutadobe/pressroom/pressreleases/201010/10 1310AdobeMobileExperienceSurvey.html

CAB. (2008). *Policy regarding the broadcast of hits by FM radio stations.* Ottawa, Canada: Canadian Association of Broadcasters. Retrieved

November 27, 2010, from
http://support.crtc.gc.ca/applicant/docs.aspx?pn_ph_no=pb2008-
1&call_id=67353&lang=E&defaultName=CAB/ACR&replyonly=&ad
dtInfo=&addtCmmt=&fnlSub

CRTC. (2009). CRTC *Communications monitoring report 2009*. Ottawa,
Canada: Canadian Radio-television and Telecommunications
Commission. Retrieved December 1 2010, from
http://www.crtc.gc.ca/eng/publications/reports/policymonitoring/2009/
2009MonitoringReportFinalEn.pdf

—. (2010). *CRTC Communications monitoring report 2010*. Ottawa,
Canada: Canadian Radio-television and Telecommunications
Commission. Retrieved November 27, 2010, from
http://www.crtc.gc.ca/eng/publications/reports/PolicyMonitoring/2010/
cmr41.htm#n21

Leggatt, H. (2010a). App-pocalypse for the mobile app? *BizReport*.
Retrieved November 4, 2010, from
http://www.bizreport.com/2010/11/app-pocolypse-for-the-mobile-
app.html#

—. (2010b). dotMobi: 2000 per cent growth in number of mobile-ready
websites. *BizReport*. Retrieved October 21, 2010, from
http://www.bizreport.com/2010/10/dotmobi-2000-growth-in-number-
of-mobile-ready-websites.html

Online ad spend continues double-digit growth. (6 December 2010).
eMarketer. Retrieved December 6, 2010, from
http://www.emarketer.com/Article.aspx?R=1008087

Slutsky, I. (2010, August 30). Chief listeners use technology to track, sort
company mentions. *AdAge Digital*. Retrieved September 4, 2010, from
http://adage.com/digital/article?article_id=145618

Statistics Canada. (2006). *2006 Census: The evolving linguistic portrait*.
Canada: Statistics Canada. Retrieved December 2 2010, from
http://www12.statcan.ca/census-recensement/2006/as-sa/97-555/p6-
eng.cfm

Acknowledgements

The author wishes to acknowledge the assistance provided by Teilhard
Gentillon during the research phase of this chapter.

CONTRIBUTORS

Pierre C. Bélanger is Professor in the Department of Communications and the Institute of Canadian Studies at the University of Ottawa, Canada. He specializes in Canadian media and telecommunications industries and in the sociology of technological innovations. From 1998 to 2001, he was on secondment to CBC/Radio-Canada where he worked as Director of New Media for Radio-Canada. Over the past six years, he has been associated with Astral Radio where he is acting as advisor on emerging technologies.

Sam Coley is Degree Leader Radio and Senior Lecturer at the Birmingham School of Media, Birmingham City University, where he teaches radio production, documentary making, commercial production and digital editing skills. He has previously worked as Creative Director for the Northern Region of The Radio Network in New Zealand and as a freelance documentary producer for Radio New Zealand. Sam has also worked as a radio trainer for the BBC World Service Trust, and as a media consultant and trainer for CARE International and the United Kingdom Prison Radio Association. In 2006, 2008 and 2011 Sam travelled across Africa to work on various audio and research projects at university radio stations, in educational radio programmes as well as designing a campaign to promote sexual health and reproduction awareness for young Ethiopians.

Harry Criticos is a PhD candidate at the University of Newcastle, Australia, in the School of Design, Communication and IT. He worked in radio as an announcer and producer for twenty-four years and became interested in this research project while working for the Super Radio Network (SRN). Having worked in regional radio, he was curious to know what effect, if any, the growth of networking may have on the industry. Harry has also presented a paper at the ANZCA 2010 conference on networking and localism in regional radio.

Janey Gordon is a Principal Lecturer in Media and the project co-ordinator for the community radio station, Radio LaB, at the University of Bedfordshire in the United Kingdom. She teaches radio broadcasting and her research interests and publications are in the areas of community radio, mobile phones and media pedagogy. She has a background as a professional radio broadcaster and started in radio as a BBC studio manager before going on to produce in schools radio and then into BBC local radio. Her first book *The RSL, Ultra-local Radio* is used at both undergraduate and postgraduate level and came about from working with the students on the University radio station. Her radio research and publications comparing Australian and UK community radio stations was funded by the British Academy. This work led to a commission from the BBC World Service Trust concerning plans for community radio in Georgia. Janey has also edited a second book, *Notions of Community, a Collection of Community Media Debates and Dilemmas*, which draws together the converging technologies of community media.

Matt Grimes is Degree Leader for Music Industries and Lecturer in Music Industries, Radio Production and Media Theory at Birmingham School of Media, Birmingham City University, UK. He is also a member of the Birmingham Centre for Media and Cultural research where he is currently studying for a PhD conducting research into British anarcho-punk, the punk canon, fandom and popular/cultural memory. Matt has facilitated various community radio projects with marginalised groups and presented papers about radio and music at numerous conferences and has recently contributed a chapter entitled "Call it Crass, but 'There Is No Authority But Yourself': Re-canonizing Punk's Underbelly" for a forthcoming publication *Sights and Sounds: Interrogating The Music Documentary* co edited by Prof D.Sanjek and Dr B.Halligan of Salford University, UK.

Peter Hoar is a Senior Lecturer in radio at Auckland University of Technology. He contributes reviews and documentaries to Radio New Zealand Concert and has worked in television, journalism and as a librarian. His research interests are in radio history and audio cultures.

Anne F. MacLennan is an assistant professor in the Department of Communication Studies at York University and the York-Ryerson Joint Graduate Program in Communication & Culture. In 2011 Anne published "Cultural Imperialism of the North? The Expansion of CBC's Northern Service and Community Radio" in *Radio Journal: International Studies in Broadcast & Audio Media*. "Resistance to Regulation: Early Canadian Broadcaster and Listeners," appeared in *Islands of Resistance: Pirate Radio in Canada*. Other publications include "Women, Radio and the Depression: a "Captive" Audience from Household Hints to Story Time and Serials" in *Women's Studies: An Interdisciplinary Journal* as well as other work on early Canadian radio programming and the audience.

Tom Morton is Associate Professor of Journalism at the University of Technology Sydney and Director of the Australian Centre for Independent Journalism. Prior to joining the academic staff at UTS he was a radio journalist and feature and documentary producer with the Australian Broadcasting Corporation for 22 years. He has a PhD in German Language and Literature from the University of Adelaide and is a member of The Lovely Assistants, Tomstu and the Garages of Desire.

J. Mark Percival is Programme Leader for Media at Queen Margaret University, Edinburgh and lives in Glasgow, Scotland. His 2007 doctoral thesis at the University of Stirling, Making Music Radio, focused on the social dynamics of the relationship between record industry pluggers and music radio programmers in the UK. Part of that work appears in an article in *Popular Music and Society* (2011). He has written about Scottish indie music production in *Popular Music History* (2009) and in Studies in Music from the University of Western Ontario (2011), and has contributed book chapters on popular music and identity to *Tartan to Tartanry: Scottish Culture, History and Myth*, and to *Britpop and the English Music Tradition*. Since 2008 he has been Chair of the UK and Ireland branch of the International Association for the Study of Popular Music (IASPM UK/I). Alongside his academic career Mark has been a Mercury Music Prize judging committee member (1998 and 1999) and a DJ for BBC Radio Scotland (1988-2000), playing alternative, indie and electronica.

Richard Rudin is Senior Lecturer in Journalism at Liverpool John Moores University in the UK. His solo-authored book, *Broadcasting in the 21ˢᵗ Century*, was published in 2011 by Palgrave Macmillan. He co-authored *An Introduction to Journalism* (Focal Press, 2002) and was one of the main contributors to the multi-volume *Encyclopedia of Radio* (Fitzroy Dearborn, 2004). He was Chair of the International Division of the Broadcast Education Association 2007-11.

Brent Simpson is a trustee and presenter for Waiheke Community Radio in New Zealand. He recently completed a Masters in the Department of Film, Television, and Media Studies at the University of Auckland on the subject of Low Power FM and community radio in New Zealand.

Siobhan Stevenson is a Lecturer in Radio Studies and Professional Development at Birmingham City University's School of Media. After working in radio production at the BBC for several years, Siobhan moved into academia full-time, two years ago. She is also a member of the Birmingham Centre for Media and Cultural Research and currently undertaking PhD research into how discourses of 'social gain' are deployed and articulated at different stages in the processes from policy to practice in British community radio in the 21st Century. Siobhan's other research interests include radio programming for diaspora communities and public service broadcasting.

Tony Stoller was Chief Executive of the UK's Radio Authority until 2003 and is now a media historian, as well as Chair of the Joseph Rowntree Foundation. His book *Sounds of Your Life* is the definitive history of Independent Radio in the UK. Tony is currently undertaking a PhD within Bournemouth University's media school in England, studying classical music on UK radio between 1945 and 1995.

Helen Wolfenden is a broadcaster and academic. She has worked for ABC Local Radio as a presenter, producer and manager, and BBC Radio Scotland as a presenter and researcher. Helen is currently a Lecturer in Radio Journalism for the University of Salford, Manchester.